→INTRODUCING

STATISTICS

EILEEN MAGNELLO & BORIN VAN LOON

ICON

This edition published in
the UK and the USA
in 2009 by Icon Books Ltd,
Omnibus Business Centre,
39–41 North Road, London N7 9DP
email: info@iconbooks.net
www.introducingbooks.com

Sold in the UK, Europe and Asia
by Faber & Faber Ltd,
Bloomsbury House,
74–77 Great Russell Street,
London WC1B 3DA or their agents

Distributed in South Africa
by Book Promotions,
Office B4, The District,
41 Sir Lowry Road,
Woodstock 7925

Distributed in Australia and
New Zealand by
Allen & Unwin Pty Ltd,
PO Box 8500,
83 Alexander Street,
Crows Nest, NSW 2065

Distributed to the trade in the USA
by Consortium Book Sales
and Distribution
The Keg House,
34 Thirteenth Avenue NE, Suite 101,
Minneapolis, MN 55413-1007

Distributed in Canada
by Penguin Books Canada,
90 Eglinton Avenue East,
Suite 700, Toronto,
Ontario M4P 2Y3

ISBN: 978-184831-056-8

Originating editor: Duncan Heath

Printed by Gutenberg Press, Malta

Drowning by Numbers

We are drowning in statistics. And they are not just numbers. For the media, statistics are routinely "damning", "horrifying", "deadly", "troublesome" – or, on occasion, "encouraging". The press constantly suggest that statistical information about crime, disease, poverty and transport delays is not only the source of the problem, but that it represents real entities or real people instead of one point on a graph.

This tendency to assign meaning to a single essence or example by looking at one point on a statistical distribution creates unnecessary confusion and fear.

Averages or Variation?
Much of the shock-horror statistical information used by the media is based on statistical **averages**. Despite the often misleading preoccupation with averages, the most important statistical concept neglected by journalists and news reporters is **variation**. This concept is essential to modern mathematical statistics and plays a pivotal role in biological, medical, educational and industrial statistics.

So why is variation important?

Variation measures individual differences, while averages are concerned with summarising this information into one exemplar.

4

Variation can be quite easily seen in multicultural Britain, and especially London, which now consists of more than 300 sub-cultures with as many languages spoken (from Acholi to Zulu) and thirteen different faiths. For some, multiculturalism is about valuing everybody and not making everyone the same (or not reducing this ethnically diverse group of individuals to one representative person).

There are so many individual differences across the British population that it is now practically meaningless to talk about the 'average' British person, as one might have done before 1950.

These multifarious individual differences embody the statistical variation that is the crux of modern mathematical statistics.

Why Study Statistics?

Statistics are used by scientists, economists, government officials, industry and manufacturers. Statistical decisions are made constantly and affect our daily lives – from the medicine we take, the treatments we receive, the aptitude and psychometric tests employers give routinely, the cars we drive, the clothes we wear (wool manufacturers use statistical tests to determine the thread weave for our comfort) to the food we eat and even the beer we drink.

Knowledge of some basic statistics can even save or extend lives – as it did for Stephen Jay Gould, whom we will hear more about later.

What are Statistics?

Yet for all their ubiquity, we don't really know what to make of statistics. As one columnist put it, "cigarettes are the biggest single cause of statistics". People express a wish to avoid bad things by saying, "I don't want to be another statistic". Do statisticians really think that all of humanity is reducible to a few numbers?

Although some people think that statistical results are irrefutable, others believe that all statistical information is deceptive.

Though Twain mistakenly attributed this aphorism to Prime Minister Benjamin Disraeli in 1904, Leonard Henry Courtney had first used the phrase in a speech in Saratoga Springs, New York in 1895, concerning proportional representation of the 44 American states.

Some government officials even blame statistics for causing economic problems. When White House press secretary Scott McClellan tried to explain in February 2004 why the Bush administration reneged on a forecast that should have led to more jobs in America, his defence was simple.

In Britain, the Statistics Commission called for "Cabinet Ministers to be banned from examining statistical information before it is made public, as this would avoid political influence or exploitation". Nevertheless, the statistics that are available for public consumption can shape public opinions, influence government policies and inform (or misinform) citizens of medical and scientific discoveries and breakthroughs.

What Does Statistics Mean?

The word "statistics" is derived from the Latin *status*, which led to the Italian word *statista,* first used in the 16th century, referring to a statist or statesman – someone concerned with matters of the state. The Germans used *Statistik* around 1750, the French introduced *statistique* in 1785 and the Dutch adopted *statistiek* in 1807.

Early statistics was a quantitative system for describing matters of state – a form of "political arithmetic".

The system was first used in 17th-century England by the London merchant **John Graunt** (1620–74) and the Irish natural philosopher **William Petty** (1623–87).

Graunt

Petty

In the 18th century, many statists were jurists; their background was often in public law (the branch of law concerned with the state itself).

It was the Scottish landowner and first president of the Board of Agriculture, **Sir John Sinclair** (1754–1834), who introduced the word "statistics" into the English language in 1798 in his *Statistical Account of Scotland*.

QUANTUM OF HAPPINESS

I wanted to measure the "quantum of happiness" of the Scots.

THE WHAT?

Sinclair used statistics for social phenomena rather than for political matters. This led eventually to the development of vital statistics in the mid-19th century.

Vital Statistics vs. Mathematical Statistics

Not all statistics are the same. There are two types: vital statistics and mathematical statistics.

Vital statistics is what most people understand by statistics. It is used as a plural noun and refers to an aggregate set of data.

It refers to the description and enumeration used in census counts or in the tabulation of official statistics such as marriage, divorce and crime statistics.

We also have insurance statistics and even cricket and baseball statistics.

This process is primarily concerned with average values, and uses life tables, percentages, proportions and ratios: probability is most commonly used for actuarial (i.e. life-insurance) purposes. It was not until the 20th century that the singular form "statistic", signifying an individual fact, came into use.

Mathematical statistics is used as a singular noun, and it arose out of the mathematical theory of probability in the late 18th century from the work of such continental mathematicians as Jacob Bernoulli, Abraham DeMoivre, Pierre-Simon Laplace and Carl Friedrich Gauss.

In the late 19th century, mathematical statistics began to take shape as a fully-fledged academic discipline in the work of **Francis Ysidro Edgeworth** (1845–1926), **John Venn** (1834–1923), **Francis Galton** (1822–1911), **W.F.R. Weldon** (1860–1906) and **Karl Pearson** (1857–1936).

We three began to apply Charles Darwin's ideas to the measurement of biological variation, which required a new statistical methodology.

Mathematical statistics encompasses a scientific discipline that analyses variation, and is often underpinned by matrix algebra. It deals with the collection, classification, description and interpretation of data from social surveys, scientific experiments and clinical trials. Probability is used for statistical tests of significance.

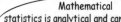

Mathematical statistics is analytical and can be used to make statistical predictions or inferences about the population.

Furthermore, it capitalizes on all the individual differences in a group by examining the spread of this statistical variation through such methods as the range and standard deviation, which we'll look at later.

Vital statistics is concerned with averages, whereas mathematical statistics deals with variation.

Used in this sense, statistics is a technical discipline, and while it is mathematical, it is essential to understand the statistical concepts underlying the mathematical procedures.

The Philosophy of Statistics

The decision to examine averages or to measure variation is rooted in philosophical ideologies that governed the thinking of statisticians, natural philosophers and scientists throughout the 19th century. The emphasis on statistical averages was underpinned by the philosophical tenets of **determinism** and **typological** ideas of biological species, which helped to perpetuate the idea of an idealized mean.

Determinism implies that there is order and perfection in the universe ...

Thus, variation is flawed, a source of error that should be eradicated, since it interferes with God's plan and purpose for His world.

The typological concept of species, which was the dominant thinking of taxonomists,* typologists and morphologists until the end of the 19th century, gave rise to the morphological concept of *species*. Species were thought to have represented an ideal type.

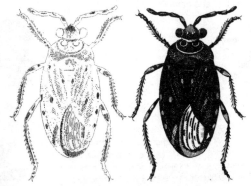

The presence of an ideal type was inferred from some sort of morphological similarity, which became the species criterion for typologists. This could have had the effect of creating a proliferation of species since any deviation from the type would have led to the classification of a new species.

Genuine change, according to the morphological concept of species was possible only through the **saltational** origins of new species, meaning that new species should have occurred by leaps or jumps in a single generation. Because Darwin's theory of evolution depended upon "gradual" changes, it was incompatible with essentialism.

Darwin

---

* **Taxonomists** classify organisms into groups
  Typologists classify organisms according to general types
  Morphologists study the forms of organisms

Darwin and Statistical Populations

The transition to measuring statistical variation represented an ideological shift that occurred during the middle of the 19th century, when **Charles Darwin** (1809–82) began to study minute biological variation in plants and animals.

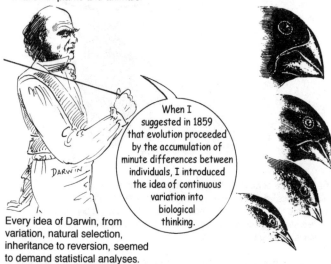

When I suggested in 1859 that evolution proceeded by the accumulation of minute differences between individuals, I introduced the idea of continuous variation into biological thinking.

DARWIN

Every idea of Darwin, from variation, natural selection, inheritance to reversion, seemed to demand statistical analyses.

Darwin had not only shown that variation was measurable and meaningful by emphasizing statistical populations rather than focusing on one type or essence, but he also discussed various types of correlation that could be used to explain natural selection. As the evolutionary biologist **Sewall Wright** (1899–1988) remarked in 1931:

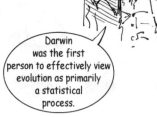

Darwin was the first person to effectively view evolution as primarily a statistical process.

Victorian Values

Although some developments in vital and mathematical statistics took place on the Continent, we owe the rapid growth and application of vital statistics in the mid-19th century and mathematical statistics in the late 19th and early 20th centuries to these Victorians:

Vital Statisticians

Edwin Chadwick
William Farr
Florence Nightingale
Thomas Rowe Edmunds

Mathematical Statisticians

Francis Ysidro Edgeworth
Francis Galton
W.F.R. Weldon
George Udny Yule
William Sealy Gosset
Karl Pearson

The development of both types of statistics took place in the wider context of the Victorian culture of measurement. The Victorians valued the precision and accuracy that instruments provided because it gave them more reliable information. In the expanding industrial economy, it was essential to establish that the results were reproducible for an international market.

Engineers and physicists spent long hours in laboratories recording and measuring electrical, mechanical and physical constants for machines, apparatus and other objects. Biologists and geologists collected as much information as possible on expeditions to create geographical maps, measure longitude and latitude, and classify new species of plants and animals.

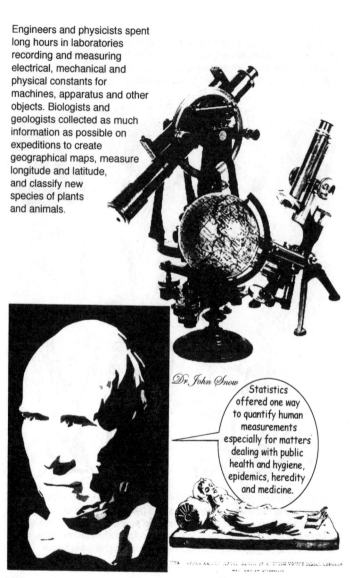

Dr John Snow

Statistics offered one way to quantify human measurements especially for matters dealing with public health and hygiene, epidemics, heredity and medicine.

Where Did it All Begin?

Counting people or taking censuses represents one of the oldest uses of statistics known to mankind: the Babylonians, Egyptians and Chinese all collected statistical information about their people, mainly to determine the number who were liable for military service and to fix the rates of taxation. In the millennium before Christ, the Romans and the Greeks were conducting censuses. The word "census" is derived from the Roman Censors, whose duty was to count their people. Their census was a register of Roman citizens and their property.

Scandinavian countries introduced the first national census in the middle of the 17th century. The United States held their first census in 1790 to ensure proportional representation in the election of Congressmen in the first thirteen American states.

Eleven years later, in 1801, an official annual census count was introduced in Britain.

Names	No. of Families	No. of Males	No. of Females				Total No. of Persons
George Fagean	1	1	3			4	4
James Dimangy	1	1	1		1	1	2
Joseph Nowatt	1	1	11			12	12
Paw Pooser Timbl.							
Maria Ker	1	1	3			4	4
James Potter	1	5	3	2		6	8
Catherine Saulthorpe	2		5			5	5
Sarah Fransted	1	3	59			62	62
Mip Pease Co	1	1	4		1	4	5
Nicholas George	1	4	6			10	10
John Berkes							
Ann Pro-ver	2		6			6	6

Parish Registers

How were people counted before a national census was introduced? Parish registers provided valuable information for some of the earliest ideas on statistical populations. The French implemented the custom of registering deaths and marriages in Burgundy in the early 14th century, and by the 16th century the registration of baptisms, marriages and deaths had become obligatory for French curates. In England, this registration became the responsibility of the local clergy in 1538, thanks to Thomas Cromwell, Lord Chancellor to Henry VIII.

I instructed the clergy of every parish to keep registers of all baptisms, weddings and funerals at which they officiated.

But dissenters and members of other faiths were largely excluded from the record, as were many within the established Church who did not wish or could not afford to pay the fees for ecclesiastical registration.

The *London Bills of Mortality*

During the 17th and 18th centuries in England, a growing number of people supported non-conformist religion instead of the established Church. Although registers were kept by Jews, Quakers and many of the Free Churches and Chapels, these were regarded as an unacceptable source of national records since they were outside the established system.

With so many people not being counted, there was a growing interest in the question of whether the country's population was increasing or decreasing.

John Graunt made one of the first systematic attempts to use 10,000 parish records in England and Wales, which contained information on sex, age and cause of death, in his *Natural and Political Observations upon the London Bills of Mortality*. Graunt used the term "political arithmetick" to describe his work – a term coined for him by his friend William Petty.

 A generall Bill for this present year, ending the 19 of *December* 1665. according to the Report made to the KINGS most Excellent Majesty.

By the Company of Parish Clerks of London, &c.

The Diseases and Casualties this year.

The cover of the book reads:
London's Dreadful visitation:
On a collection of all the

Bills of Mortality

For the Present Year:
Beginning the 17th of December 1664, and
ending the 19th December following
As also the GENERAL or whole year bill
According to the Report made to the
King's Most Excellent Majesty
By the company of Parish Clerks, London

Halley's Mortality Tables

The exceptional work on mortality data in the 18th century was the construction of the Life Table (sometimes referred to as the "Table of Vitality"). This was first suggested by John Graunt and conceived by the astronomer **Edmond Halley** (1656–1742), best known for his eponymous comet.

I produced the first table of cometary elements in 1676 and constructed the first scientific mortality table in 1693.

The Dutch astronomer and political arithmetician **Nicolaas Struyck** (1687–1769) built on Halley's work on comets and his research into determining the size of various populations. Struyck organized large-scale population surveys in the Netherlands, but his greatest ambition was to make a reasoned estimate of the total number of people on earth. He wanted to know if the population was increasing, stable or decreasing.

22

Malthusian Populations

While various commentators tried to determine the population of a country or the world, the economist **Thomas Robert Malthus** (1766–1834) argued in his celebrated work *An Essay on the Principle of Population* (1798) that unchecked human populations would always exceed the means of subsistence and that human improvement depended on stern limits on reproduction.

Malthus believed that populations would increase exponentially (2, 4, 8, 16, 32, etc.), whereas food supplies would increase mathematically (2, 4, 6, 8, 10, etc.). Malthus' hypothesis implied that the actual population would always have a tendency to push above the food supply.

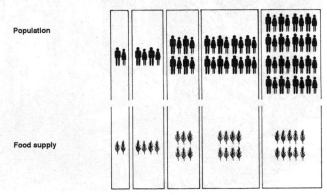

Demography – the Science of Populations

Any attempt to improve the conditions of the lower classes by increasing their incomes or boosting agricultural productivity was seen by Malthus as fruitless. He believed that "moral restraint" was needed to reduce the growth rate. **Demography** began as the numerical study of poverty.

While Malthus had considered how population growth might constrain prosperity in the late 18th century, it wasn't until the mid-19th century that the collection of population statistics in Europe and the US had become extensive enough to contemplate a *science of populations*. The grandfather of the Bertillion family of French demographers, **Jean Paul Achille Guillard** (1799–1896), first used the word *démographie* in 1855 for this new science.

Demography deals with the size, conditions, structure and movement of populations as well as the vital statistics of birth, marriage and death to describe those populations.

Competition between France and England, sharpened by the French Revolution and the descent of Europe into war after 1793, led English society to consider its military manpower and population resources in the last decade of the 18th century.

During the Napoleonic War years, the utilitarian philosopher **Jeremy Bentham** (1748–1832) discovered that the government didn't know how many paupers received relief, and that it couldn't even account for the amount of money in circulation.

Bentham

This lack of essential information indicates a basic instability in the affairs of the state and points to the need for a national system of record-keeping.

The Statistical Society of London

The lack of official registration was the driving force in the formation of the Statistical Society of London (now the Royal Statistical Society) in 1834. Malthus, together with the Belgian statistician and meteorologist **Adolphe Quetelet** (1796–1874) and **Charles Babbage** (1791–1871), who developed the universal calculating machine (a precursor of the computer), joined forces to establish the Society.

Our first recommendation was for a national registration system with a central office in London.

Following legislation in 1836, civil registration was introduced, requiring the notification of births, marriages and deaths.

The General Register Office (GRO) was established, giving England and Wales a system of demographic recording unique in Europe at that time. The first comprehensive census was not undertaken in England until 1851, when provisions were made to include age, sex, occupation and birthplace, as well as counting the blind and the deaf.

Edwin Chadwick and Sanitary Reforms

The first census produced detailed information about the number of deaths from disease, and led to an awareness of the appalling sanitary conditions in towns. Overcrowding often led to inadequate housing without proper ventilation and sanitation. Cesspools overflowed and sewers ran directly into rivers, creating major health risks for everyone.

A key figure in sanitary reform and in the use of statistics was the liberal-minded **Edwin Chadwick** (1800–90), who was involved in the government's reorganization of aid to the poor and destitute.

The success of the sanitary reforms gave new importance to the collection of statistics.

When the chief question on the conditions of England became the sanitary question, the function of statistics became the measurement of health.

27

William Farr and Vital Statistics

After the GRO was set up, Chadwick recommended the appointment of a Registrar General for the registration of births and deaths. The post was created by Parliament, and **Thomas Henry Lister** (1800–42), the brother-in-law of the Secretary of State, who was known to various Ministers, was appointed.

However, there was a requirement to compile the statistical records, and Lister recruited **William Farr** (1807–93) to analyse the statistics, as he was the only medical man who paid any attention to vital statistics.

Farr's work as Statistical Superintendent at the General Register in 1839 was a landmark in the development of English preventive medicine and medical statistics. His methods and organization of vital statistics provided a template for all nations. Together with **Thomas Rowe Edmunds** (1803–99) they created the modern discipline of vital statistics

Florence Nightingale: the Passionate Statistician

The statistical work of Farr and Quetelet provided inspiration for **Florence Nightingale** (1820–1910), one of the most famous and easily recognizable Victorians, known to everyone as the "Lady with the Lamp" who made nursing a respectable profession. Yet we know little about her role as the "Passionate Statistician", the sobriquet given to her in 1913 by her first biographer, Sir Edward Cook.

Yet in my capacity as a statistician I was able to introduce essential measures of sanitary reform in hospitals in the battlefield and in London.

By using the methods and ideas of the mid-Victorian statisticians, Nightingale persuaded various government officials of the importance of the lessons she learned in the Crimean War, and showed that mortality rates could be reduced among the Army at home.

29

As a young woman, Florence met a number of Victorian scientists at dinner parties, including Charles Babbage. She was so fascinated with numbers at an early age that by the time she was twenty she was receiving two-hour instructions from the Cambridge mathematician **J.J. Sylvester** (1814–97).

In the mornings Florence would study material on the statistics of public health and hospitals, accumulating a formidable array of statistical information. Her enjoyment was so immense that she found "the sight of a long column of figures was perfectly reviving".

Statistics is the most important science in the world. To understand God's thought, we must study statistics for these are the measure of His purpose.

She shared with Francis Galton the idea that the statistical study of natural phenomena was the "religious duty of man".

3705
0,352
189
353
478
987
23
4679
9573
266
493
4,870
34
48
93
844
28
84
33,00
380
517
3. 4

The Statistics of the Crimean War
In 1854, Nightingale's lifelong friend, the Secretary at War **Sidney Herbert** (1810–61), approached her with an offer.

Herbert

I asked her to be "Superintendent of the female nursing establishment in the English General Military Hospitals in Turkey".

She was to care for the British troops fighting in the Crimean War, and was to take a group of 38 nurses with her.

Her connections with the government and her years of advocacy for professional nursing gained her the prestige that made possible this exceptional appointment. Prior to this time, women had never been allowed to serve officially.

Herbert had responded to public outrage at reports in *The Times* ...

... reports of the suffering of the common soldiers like us, caused by the incompetence of the Army commanders.

Herbert hoped that Nightingale's presence could pacify the public. Readers of *The Times* donated £7,000 for her personal use, which was eventually used to improve hospital conditions, but it also inspired jealousy among Army doctors and officers.

Once Nightingale arrived in the Crimea, she found herself amid utter chaos in the hospital at Scutari: there was no furniture, food, cooking utensils, blankets or beds; rats and fleas were constant problems. Though she was able to get basins of milkless tea from the hospital, the same basin was used by the soldiers for washing, eating and drinking.

She was the only person with funds and the authority to rectify this bleak situation. She requested eating utensils, shirts, sheets, blankets, stuffed bags for mattresses, operating tables, screens and clean linen. She soon set up a laundry and a kitchen, and much of the food was supplied by Fortnum & Mason.

I was on my feet constantly, especially as I was the only nurse allowed on the wards after 8pm.

We called her the "Lady with the Lamp".

Mortality Statistics in the Crimea

Nightingale was very distressed by the statistical carelessness she found in the military hospitals. There was a complete lack of coordination among hospitals, and no standardized or consistent reporting. Each hospital used its own classification of disease, tabulated on different forms, making comparisons impossible. Even the number of deaths was not accurate: hundreds of men had been buried, but their deaths were not recorded.

Polar Area Graphs

Although various 19th-century vital statisticians used an assortment of graphs and tables for their statistical results, Nightingale helped popularize the use of pictorial diagrams for statistical information. She developed her Polar Area Graph, cut into twelve equal angles: the slices represented the months of the year and revealed changes over time.

AUGUST
JULY
SEPTEMBER
JUNE
MAY
BULGARIA
APRIL 1854
CRIMEA
OCTOBER
MARCH 1855
NOVEMBER
FEBRUARY
DECEMBER
JANUARY 1855

April 1854 to March 1855

☐ Death from wounds in battle

☐ Death from other causes

▨ Death from disease

My graph not only dramatized the extent of the needless deaths during the war, but it also persuaded the medical profession that deaths were preventable if sanitation reforms were implemented in hospitals.

After the war, Florence wrote to Quetelet: "On my part this passionate study of statistics is not in the least based on a love of science, rather it comes from the facts that I have seen so much of the misery and suffering of humanity, of the irrelevance of laws and Governments."

34

Probability

How did 19th-century statisticians reduce data to something more manageable? While data was summarized in diagrams and tables, until the end of the 19th century the two main statistical tools were probability and averages.

Probability is one of the oldest statistical concepts: notions of probability were used as a tool to solve problems in games of chance beginning in the 14th century.

There are different approaches to probability:
1. **Subjective**
2. **Games of chance**
3. **Mathematical**
4. **Relative frequency**
5. **Bayesian**

With six main probability distributions:
1. **Binomial distribution**
2. **Poisson distribution**
3. **Normal distribution**
4. **Chi-square distribution**
5. **t distribution**
6. **F distribution**

The first three distributions are discussed on pages 47–50. The last three distributions are used to determine the statistical significance of the chi-square (pages 153–6). t and F tests are examined later, on pages 165 and 170 respectively.)

There are two types of statistical distributions:
probability distributions, which describe the possible events in a sample and the frequency with which each will occur;
and **frequency distributions** (see pages 74, 76 and 79–85).

Statisticians use probability distributions to interpret the results from a set of data that has been analysed by using various statistical methods. Frequency distributions transform very large groups of numbers into a more manageable form and show how frequently a particular item or unit in a group occurs.

Variables

Variables are characteristics of an individual or a system that can be measured or counted. These can vary over time or between individuals.

Variables can be classified into two groups:

Categories that can be counted are called discrete (like eye-colour, gender or political affiliation).

Quantities that can be measured are called continuous (like height, weight or blood pressure).

Discrete:
things you can tick

Eye colour:
☐ Brown
☐ Blue
☐ Green
☐ Grey

Gender:
☐ Male
☐ Female

Politics:
☐ Labour
☐ Conservative
☐ Liberal

Continuous:
things you can read off a scale

These variables can be further subdivided, and this is discussed later.

The **Subjective approach** to probability involves a degree of rational belief.

Probability is assessed through some betting procedure such as ...

What is the past form of the horse? What is the state of the turf? What is the competition like?

The possible outcomes often reflect personal opinions. Two people could use different probabilities, but there is no objective procedure to determine if one is right and the other is wrong.

Gaming theory is assessed through a betting scheme based on what the person thinks the probability of some outcome will be. The idea is to locate probability where it should be in the mind of the observer, not in the outside world. The problem is that people with equal knowledge and skills can come to different answers.

Author

Games of Chance

Games of chance have been around as long as man has been able to throw dice. From archaeological evidence in northern Iraq, man was playing such games in ancient Mesopotamia before the beginning of the 3rd millennium BC. Dice were also used at the time of the XVIIIth dynasty in Egypt (c. 1400 BC).

Primitive dice were formed by roughly squaring the long bone of an animal and cutting it into sections to create objects approximately cubical in shape. The astragalus (a small bone in the ankle) was used commonly in the gaming which the Greeks and later the Romans played.

I discussed the different throws that could be made with three dice in my 'Divine Comedy'.

Dante Alighieri
(1265–1321)

The Italian Renaissance physician and mathematician **Girolamo Cardano** (1501–76) was an inveterate gambler who often had to support himself through these means. He produced the first recorded work on probability: *Liber de Ludo Aleae* (The Book on Games of Chance), published posthumously in 1633, which was used as a manual for gamblers.

In my calculations, I also appealed to the intervention of luck.

But luck was banished in the 17th century, when classical probability theory arrived. The classical theory stipulated that a full range of probabilistic events had to be anchored to mathematical probability. Thus, even Cardano's fortuitous events had to meet these mathematical requirements.

De Moivre and Gambling in Soho

The French mathematician **Abraham de Moivre** (1667–1754) wrote the *Doctrine of Chance: or A Method of Calculating the Probabilities of Events in Play* in 1718, which was based on problems concerning the advantages of players and the size of their wagers in games of chance. Like Cardano's work, de Moivre's book was also used as a gambler's manual.

He left France for England in 1685, due to Louis IV's revocation of the Edict of Nantes. This ended toleration of Protestants in Catholic France, causing hundreds of thousands of them to flee.

When he was in London, de Moivre socialized with Edmond Halley and Isaac Newton, and he was elected a Fellow of the Royal Society when he was 30 years old.

The Mathematical Theory of Probability

By the end of the 17th century, probabilistic ideas on the mathematics of permutations and combinations had been applied to games of chance by:

Christiaan Huygens (1625-1695)

Pierre de Fermat (1601?-1665)

Blaise Pascal (1623-1662)

Gottfried Wilhelm Leibnitz (1646-1716)

John Arbuthnot (1667-1735)

... but they did not know how to quantify this uncertainty.

The mathematical theory of probability gave statisticians a tool to reduce mathematical complexity, to show how regularity in a set of data could have developed out of chance and that even chance could be reduced to a set of laws.

This approach describes the long-term regularity in random events, and is the ratio of the number of favourable occurrences of events to...

$$\frac{successful\ ways}{possible\ ways}$$

It is a theoretical approach which does not involve examining real objects – one simply sets up hypothetical conditions and then works out the probability by using the binomial distribution (see pages 47–8).

Thus, one can hypothesize that a coin is unbiased (meaning that it comes up equally heads or tails) ...

... and figure out what is the probability of getting various combinations of heads and tails in many flips of the coin.

This mathematical development materialized in the 17th century, emerging as a formal theory at the beginning of the 18th, though the application of probability to statistical theory didn't take place until the end of the 19th century.

Relative Frequency

Relative frequency is an approach that makes it possible to make formal probability statements (P, A) about uncertain events, where "P" is the probability of an uncertain event "A". Thus, the probability of an event happening is the proportion of times that events of the same kind will appear in the long run.

On time *Late*

For example, if aeroplanes arrive on time in 80% of their flights, the probability of arriving on time is .80.

Probability of arriving on time = .80

This is a more scientific and objective approach than the other types of probability, and is used for finding out about the world and assessing actual existing objects. One can flip a coin 100 times and record the number of heads and tails and the ratio of the number of heads to the total number of flips.

In one of his earliest lectures on statistics, Karl Pearson scattered hundreds of pennies all over the lecture room floor and asked his students to pick them up and arrange them in heads or tails.

The result was very nearly half heads and half tails, thus proving the law of averages and probability.

But how do we know how many times to flip a coin (or roll a die) in order for it to be an adequate test? If you flip a coin and get 60 heads and 40 tails, you wouldn't get the same results the next time. Probability will always change, and by the time a stable probability is reached the coin may be worn out.

The way around this situation is the Relative Frequency Ratio, which is the ratio of the number of times that an event occurs in a series of experimental trials divided by the number of actual trials in the experiment performed.

44

The Bayesian Approach

The mathematician Reverend **Thomas Bayes** (1702–61) first used probability inductively, establishing a mathematical basis for probability inference. However, the term "Bayesian" as applied to statistics was not introduced until some time around 1950.

Bayes' theorem is a formula that shows how existing beliefs, formally expressed as probability distributions, are modified by new information.

My approach is a means of calculating from the number of times that an event has **not** occurred to determine the probability that it **will** occur in future trials.

It is related to a subjective degree of belief in the induction process, and measures the plausibility of an event given incomplete knowledge.

Bayes

Example

Bayes' theorem can be used in diagnostic testing by general practitioners or clinicians. These doctors often start out with a prior belief about whether a patient has a particular illness or disease (based on the knowledge about the patient's symptoms or the prevalence of the disease in the community) and this knowledge will be modified or updated by the results of clinical tests.

45

Probability Distributions

The **Binomial distribution** is a discrete probability distribution and represents the probability of two outcomes, which may or may not occur. It describes the possible number of times that a particular event will occur in a sequence of observations. For example, it will give the probability of obtaining five tails when tossing ten coins.

EXPANDED
BINOMIAL
DISTRIBUTION
OF n=10

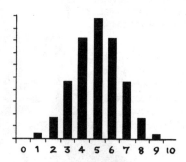

0 1 2 3 4 5 6 7 8 9 10

This distribution was introduced by the Swiss mathematician **Jacques Bernoulli** (1655–1705), whose celebrated treatise, *Ars conjectandi* (Art of Conjecturing), was published posthumously in 1713. This work represented the beginning of the mathematical theory of probability.

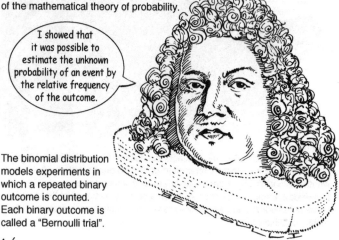

I showed that it was possible to estimate the unknown probability of an event by the relative frequency of the outcome.

The binomial distribution models experiments in which a repeated binary outcome is counted. Each binary outcome is called a "Bernoulli trial".

The binomial distribution $(p + q)^n$

is determined by the number of observations **n**,

and the probability of occurrence, denoted by **p + q** (the two possible outcomes).

This provides a model for various probabilities of outcomes that can occur. To determine the probability of each outcome, the binomial distribution has to be expanded by the number of observations – by raising **p + q** to the **n**th power.

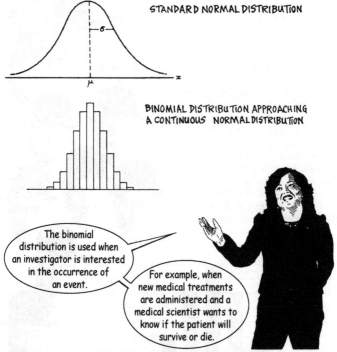

STANDARD NORMAL DISTRIBUTION

BINOMIAL DISTRIBUTION APPROACHING A CONTINUOUS NORMAL DISTRIBUTION

The binomial distribution is used when an investigator is interested in the occurrence of an event.

For example, when new medical treatments are administered and a medical scientist wants to know if the patient will survive or die.

These probability distributions correspond to different types of variables. Discrete probability distributions, such as the binomial, use discrete data (such as "heads" or "tails" in a flip of a coin) while continuous distributions, such as the normal, use continuous data (height and weight).

47

In the following example of coin-tossing, the number of observations is **n = 2** and the number of outcomes is 2 (heads or tails). To test an unbiased coin, the binomial distribution has to be expanded to accommodate the number of times the coin is flipped.

Expand the binomial distribution $(p + q)^n$ by raising p + q to the nth power (which means to multiply the number by itself).

- **p** and **q** must add up to 1
  (flipping a coin has 2 outcomes: p = ½ and q = ½)

- **n** = the number of trials or flips (2 in this example)

- The binomial distribution is $(p + q)^2$

- This is the expansion of the binomial for coin-tossing:

Suppose a coin is flipped 10 times and the result is 10 heads. The binomial distribution would account for the 10 different flips using the rules above. The chances of this would be $½^{10}$ (½ raised to the 10th power, which is 1/1024).

This means that the chances are
<u>less than 1 in 1000</u>
that an unbiased coin will produce 10 heads.

Expanded binomial distribution when n=10

0 1 2 3 4 5 6 7 8 9 10

The Poisson Distribution

The Poisson distribution, discovered by **Siméon-Denis Poisson** (1781–1840), is a discrete probability distribution used to describe the occurrence of unlikely events in a large number of independent repeated trials. The Poisson is a good approximation to the binomial distribution when the probability is small and the number of trials is large.

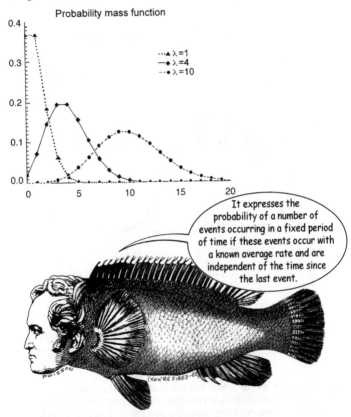

Probability mass function

It expresses the probability of a number of events occurring in a fixed period of time if these events occur with a known average rate and are independent of the time since the last event.

POISSON

(YOU'RE FIRED-E)

The analysis of mortality statistics often employs Poisson distributions on the assumption that deaths from most diseases occur independently and at random in populations.

49

The Normal Distribution

The Normal distribution is a continuous distribution, and is related to the binomial. As **n** approaches infinity, the binomial will approach the normal distribution as its limit. That is, as the binomial connects an infinite number of infinitesimal little bars the binomial will become the normal distribution.

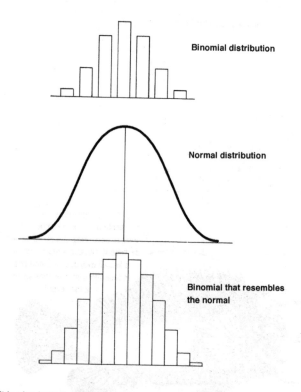

Binomial distribution

Normal distribution

Binomial that resembles the normal

It is also known as the normal curve, sometimes (inaccurately) referred to as the Gaussian distribution, and has long been used as a yardstick to compare other types of statistical distributions. It plays a vital role in modern statistics because it enables statisticians to interpret their data by using various statistical methods, which are quite often modelled on the normal distribution.

50

Astronomical Observations

The idea of a normal curve had its beginnings in calculating combinations of observations by astronomers. They used the "law of errors" (i.e. the normal curve) to combine linear equations of observations in astronomy and geodesy*.

The astronomers' methods, which were quite often ad hoc procedures with little interest in formal models of probability, required the cooperation of a group of scientists. But when mathematical statisticians began to devise statistical methods, this made it possible for one person alone to analyse the data.

De Moivre's work on games of chance and his use of the binomial theorem provided the first known derivation of the normal curve in 1733, referred to initially as the "law of error". He also assembled the first probability table for the normal distribution.

*The study of the shape and area of the earth.

51

The Central Limit Theorem

The French mathematician and astronomer **Pierre-Simon Laplace**
(1749–1827) was responsible for advancing probability as a tool for the
reduction and measurement of uncertainty in data. By 1789 he realized
that measurements were affected by a number of independent small
errors, and showed that the law of error could be derived
mathematically. Following this, he made his most important contribution
to statistics through his work on the **Central Limit Theorem** in 1810.

This was one of the most remarkable results of the theory
of probability …

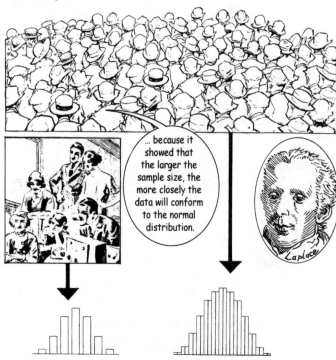

… because it
showed that
the larger the
sample size, the
more closely the
data will conform
to the normal
distribution.

Or as statisticians would say: the sampling distribution of means gets
closer and closer to the normal curve as the sample size increases,
despite any departure from normality in the population distribution.

The reason that so many variables – such as stature and intelligence – are normally distributed has to do with Laplace's Central Limit Theorem.

The mathematical underpinnings of this theorem state that data which are influenced by a very large number of many small and unrelated random effects will be approximately normally distributed.

Pierre-Simon Laplace

The Gaussian Curve and the Principle of Least Squares

Laplace's work remained the most influential book on mathematical probability until the end of the 19th century, when **Carl Friedrich Gauss** (1777–1855) advanced Laplace's idea in explicitly probabilistic terms. The result of this work was eventually (and somewhat erroneously) named the "Gaussian curve", since it had already been discovered by Laplace.

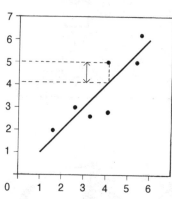

But I had already discovered this in 1805.

Adrien Marie Legendre (1752-1833)

And I reached <u>the principle of least squares</u> in 1809.

I acknowledged my debt to Laplace and made use of the law of probability in my work on the theory of the motion of heavenly bodies.

The principle of least squares, based on the theory of errors, had been devised in the early 19th century by such mathematicians and astronomers as Gauss, Laplace and Legendre to determine, for example, the shape of the Earth. It would find its greatest use in statistics at the end of the 19th century for interpreting statistical regression (see pages 128-31).

What's Normal?

Norma is Latin for a T-square, first used by masons and carpenters in antiquity to make their work rectangular. From their use of the T-square, a right angle became known as a "normal angle", a term subsequently used in geometry in the 17th century. Gauss, who studied the normal curve in 1809, used the word "norm" in algebra at the end of the 18th century.

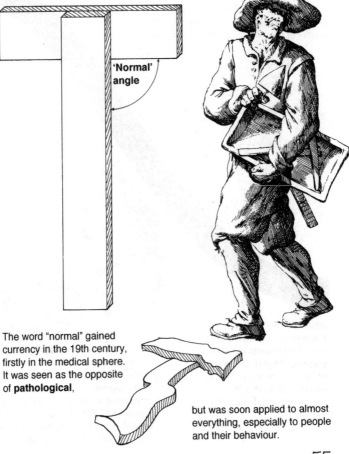

'Normal' angle

The word "normal" gained currency in the 19th century, firstly in the medical sphere. It was seen as the opposite of **pathological**,

but was soon applied to almost everything, especially to people and their behaviour.

55

"Normal" was thus used to express how things *are* or how they *ought to be*, and eventually was used to describe the bell-shaped symmetric distribution that had been used quite extensively by astronomers since the 17th century and by statisticians since the 1870s.

Yet, as **Ian Hacking** observed, the word "normal" is underpinned by a duality of meaning.

The norm may be what is usual or typical, yet our most powerful ethical constraints are also called the norm.

HACKING

STIGLER

KRUSKAL

While "normal" signifies the average or usual and the "norm" represents the ideal, **Stephen Stigler** and **William Kruskal** have shown that in statistics there is a third component that blends the first two.

This happens when statisticians refer to the asymptotic* normal limit, or the "usual limit" that is still not fully attainable.

*Asymptotic: continually approaching a given curve but not meeting it at a finite distance.

The Naming of the Normal

While Quetelet used "binomial law" to describe this distribution, Galton used "error curve" and eventually christened it the "normal curve" in February 1877, when he read his paper, "Typical Laws of Heredity", at the Royal Institution.

The American logician and mathematician **Charles Sanders Peirce** (1839–1914) and the German mathematician **Wilhelm Lexis** (1837–1914) also introduced the word independently in 1877.

I began to use "normal distribution" in my lectures in October 1893.

Karl Pearson

Once I found that the Gaussian curve was first discovered by Laplace, I proposed to call it the Laplace-Gaussian curve, and eventually referred to it as the normal curve to avoid international questions of priority.

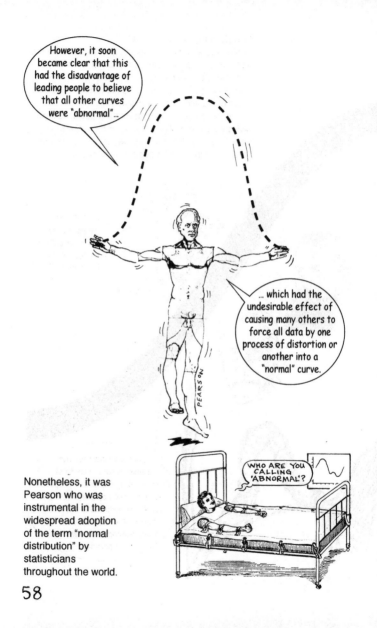

However, it soon became clear that this had the disadvantage of leading people to believe that all other curves were "abnormal"...

... which had the undesirable effect of causing many others to force all data by one process of distortion or another into a "normal" curve.

Nonetheless, it was Pearson who was instrumental in the widespread adoption of the term "normal distribution" by statisticians throughout the world.

WHO ARE YOU CALLING 'ABNORMAL'?

58

So What is the Normal Distribution?

To a statistician it is a theoretical construct used to express what might be true of the relation between data that one collects and the probability of those values occurring by chance.

The normal curve has three mathematical properties:
1. It is a bell-shaped symmetrical curve, which is continuous and ranges from negative infinity to positive infinity.

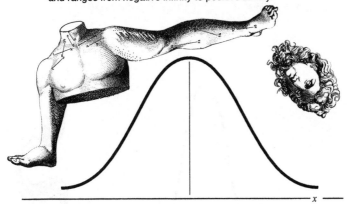

A rectangular distribution is also symmetrical because it has equal frequencies at all positions on the X axis.

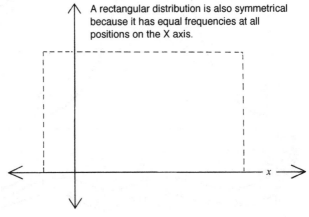

2. The mean (see pages 65–7) and the standard deviation (see pages 99–102) define its shape; the theoretical normal distribution has a population mean of zero and a standard deviation of 1. Different standard deviations will produce slightly different shapes.

The mean is the placement of the distribution on the X axis and variability, which shows how scores scatter or spread. In these figures, the mean is in the same location but curve B has more variability than curve A.

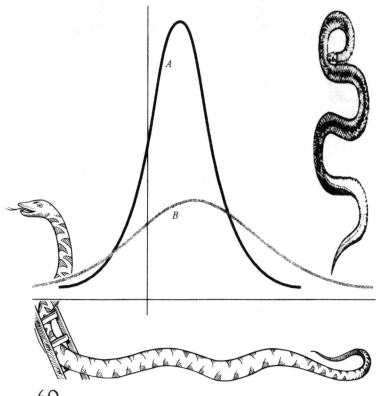

3. The *skewness* of the normal curve is zero, because it is symmetric around the mean. If the distribution were skewed to the left side, a measure of skewness would produce a negative value; if skewed to the right, this would result in positive value.

The direction of the tail indicates if it is positively or negatively skewed.

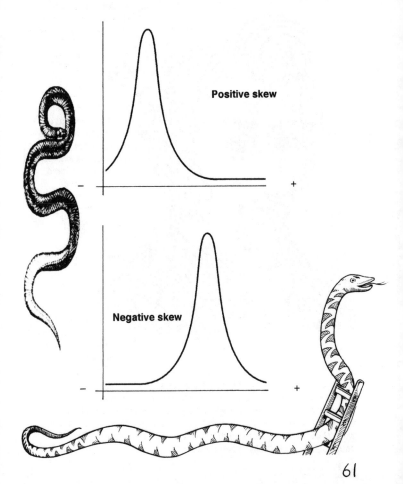

Positive skew

Negative skew

Quetelismus

The normal distribution yielded considerable power for a number of 19th-century mathematicians, philosophers and statisticians, especially Adolphe Quetelet and Francis Galton. Both believed that virtually all data had to conform to the normal curve.

Quetelet attached considerable significance to the normal curve because of his belief in determinism.

This meant that there was an ideal statistical mean value and that the normal curve was the ideal curve, since it followed the law of errors.

Hence, all the variation around the mean had to conform to this curve.

Quetelet's conviction that observational data could be fitted only to the normal curve was so strong that a doctrine was nicknamed "Quetelismus", on the grounds that he exaggerated the prevalence of the normal curve. Though Quetelet was aware that many distributions were skewed, he considered that this was "due to curious accidental causes acting unequally in two directions".

Galton's Pantograph

Inspired by Quetelet, Galton became so committed to the idea of a ubiquitous normal curve that he created a mechanical device, a modified pantograph, to stretch or squeeze any figure in two directions.

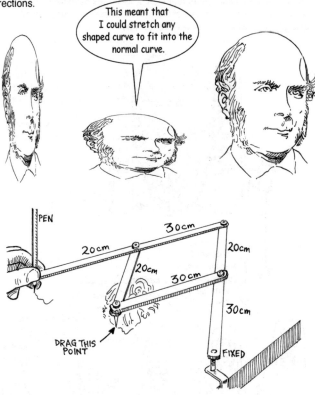

This uncompromising belief in the normal curve would become divisive between the older school of vital statistics and the new, emergent one of mathematical statistics. Such was the tyranny of the normal curve that by the end of the 19th century most statisticians assumed that no other curve could be used to describe data. But this monolithic view would be challenged by Pearson in the last decade of the century.

How to Summarise the Data?

Averages

Averages are the other principal tool of vital statisticians, and are one of the oldest statistical concepts. Ideas of averages have been used since antiquity. Aristotle spoke of the "golden mean" – where golden meant "good" – which fell between extremes.

But for statisticians there are three kinds of averages: the arithmetical mean, the median and the mode.

Quetelet and the Arithmetical Mean

This method was popularized by Quetelet in the 1830s when he discovered that astronomical error laws could be applied to the distribution of human features such as height and girth. This realization led, in turn, to his celebrated construct of *l'homme moyen* (the average man).

The regularities that Quetelet found in man and in meteors were comparable with the laws of physics. He spoke of the social system in the same way that an astronomer spoke of the system of the universe.

WE ARE STARDUST, I WE ARE GOLDEN...

AVERAGE MAN   AVERAGE MAN

I aligned the average man with the centre of gravity, referring to my work as "social physics".

I devised this phrase, but then coined the word "sociology" after Quetelet began to use "social physics".

French philosopher **Auguste Comte** (1778–1857)

Quetelet also noticed a similarity between the occurrences of regularities in phenomena in nature and in phenomena in society. He was convinced that mean values could be used to find the ideal type of society, politics and morals. Since deviations from central values caused society's ills, a mean philosophical and political position should be able to resolve society's conflicts.

As mean values are of scientific value only when they represent a type, deviations from this average are flawed and a product of error.

In 1836, Quetelet was tutoring Princes Ernest and Albert of Saxe-Coburg and Gotha, the latter to become the consort to Queen Victoria.

I was impressed by Quetelet and I later played an important role in fostering his relations with British scientists.

PRINCE ALBERT

Quetelet

The *Mean*

The **mean** is what most people are accustomed to calling an average. It involves adding up all of the values in a set of data (X) and then dividing by the total number (N) of cases.

Mean = ΣX ÷ N
Σ = summation
X = raw scores
N = total number

X
3
6
8
12
16
45 ÷ 5 = 9 = Mean

The **Median** is the point that divides the distribution into a lower half and an upper half so that 50% of the values are in one half and 50% are in the other.

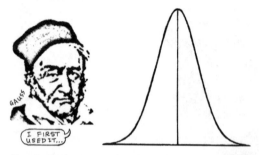

Francis Galton wanted to find a faster way to establish an average, rather than going through the trouble of calculating the mean value. He introduced the word *percentile*, which is the point that divides a distribution into a lower percentage of cases and an upper percentage.

Though Gauss first used the median in 1816, Galton introduced it into statistics. In 1874 he devised a statistical scale to find the median when he introduced the 50th percentile as the middle point in a set of data, where a set of data is divided exactly into half.

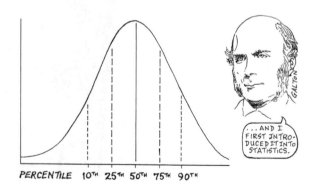

PERCENTILE   10TH   25TH   50TH   75TH   90TH

MEDIAN

The median is relatively simple to use and requires even less work than calculating the mean. When Galton wanted to measure stature in men, he aligned 100 men, from the tallest to the shortest, and picked "the one as near to the middle as may be", who represented the 50th percentile or the median.

50% ARE TALLER THAN ME.

50% ARE SHORTER THAN ME.

MEDIAN

Locating this one point took me considerably less time than finding the arithmetical mean, which involved adding up the heights of 100 men and then dividing the final number by 100.

Galton

The median is easy to find when there is an odd number of values in a set of data.

But what happens if there isn't an exact mid-point to select the median?

Group B

8
7
6
5 — — When there are two middle scores you have to calculate their means to find the median
4 — — 5+4 = 9 ÷ 2 = 4.5 — — **median**
3
2
1

Group A

7
6
5
4 — — **median**
3
2
1

SPECIMENS OF COMPOSITE PORTRAITURE
PERSONAL AND FAMILY.

HEALTH.   DISEASE.   CRIMINALITY.

CONSUMPTION AND OTHER MALADIES

Galton even devised a way to photograph the average man by superimposing many different persons that merged into one figure. He called this "composite photography".

MODE

The third measure of central tendency, the mode, quoted by Karl Pearson in 1894, is the value that occurs more frequently than any other. This finds its greatest use in advertising, which deals in concepts like the "modal family".

The mode is a point of maximum frequency; it is used most often to look for typical cases. The mode may or may not compare to an actual value. A modal family in one study might work out to 3.79 people instead of 4 people.

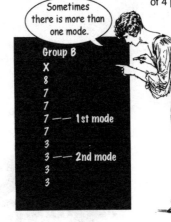

Sometimes there is more than one mode.

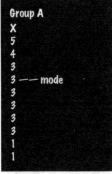

Group A		Group B	
X		X	
5		8	
4		7	
3		7	
3	— mode	7	— 1st mode
3		7	
3		3	
3		3	— 2nd mode
3		3	
3		3	
1			
1			

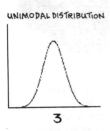

UNIMODAL DISTRIBUTION

3

BIMODAL DISTRIBUTION (WITH TWO MODES)

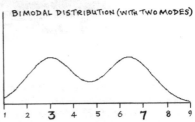

1 2 3 4 5 6 7 8 9

In Group A, there is one value that occurs six times, so the mode = 3, but in Group B there are two modes: 7 and 3. This is called bimodal.

Does it Matter Which Statistical Average is Used?

The advantage of using the mean is that the calculation is straightforward and it uses all the data in a group. However, if some items are of very low or very high values, this will distort the mean value.

And then the mean will appear to be an unrealistic indicator.

The mean is like a loaded gun, which in the inexperienced hand can lead to serious accidents, as means can give hopelessly distorted results.

Karl Pearson

The median, however, is unaffected by extreme values. For example, if one were to locate the median salary in this group of figures: £40,000; £60,000; £120,000; £160,000; £820,000 – the median would be the middle figure of £120,000. This average method would be useful to determine income, since the extreme value of £820,000 skews or distorts the data and produces an arithmetic mean value of £240,000, which is not representative of anyone's salary.

Let's consider all three measures of central tendency for calculating average salaries for a group of 41 individuals in a company.

$\boldsymbol{X}$ = One person

Number of people	Salary	
XX	£4,000	
XXXXXX	£6,000	
XXXXXXXX	£10,000	—— mode: the one occurring most frequently
XXXX	£18,000	
X	£24,000	—— median: the one in the middle with twenty people above and twenty people below
XXXX	£30,000	
XXX	£36,000	
XXXXX	£40,000	
XX	£45,000	
XXXX	£50,000	
X	£70,000	
X	£200,000	

The mean value = **£60,400**
The modal value (with eight people) = **£10,000**
The median value = **£24,000**

Misleading With Statistics

In this example, the three measures give very different figures for the average. We can immediately see that it's possible to deliberately mislead people by selecting the average that best serves one's own agenda.

For example, I could claim that my employees earn a decent wage by using the mean value of £56,524.

Yet only two people make that much money.

An investigative journalist might claim that the average (modal) salary is £10,000 and argue that half the employees earn below the average national income.

The *median* of £24,000 is perhaps the most representative, though one would get an even more realistic result if one did not include the boss's salary of £200,000, which is such an extreme value in comparison to the others. Statisticians refer to this type of extreme value as an "outlier" because it lies so far out of the distribution.

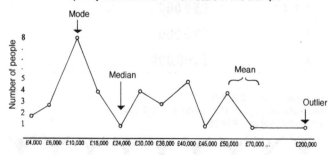

Frequency distribution of 37 individuals in this example

"The Median is Not the Message"

The only way to think about average values is to consider the full range of information, specifically the variation around the mean values: this is often a more realistic way of locating individual information.

This was the salutary lesson that the palaeontologist and evolutionary biologist Stephen Jay Gould learned shortly after he was diagnosed with mesathelioma (a rare and serious cancer usually associated with exposure to asbestos) in 1982. His knowledge of statistics helped him to realize that he did not have to be simply a statistic who would conform to the median mortality of eight months that the medical literature suggested at the time.

75

An important practical tool for depicting statistical data is a **frequency distribution** (see page 35). Gould understood that this graph did not mean he would necessarily be dead in eight months. Instead it could be interpreted to mean that he could quite easily be somewhere on the right side of the distribution where half of the patients live longer than eight months.

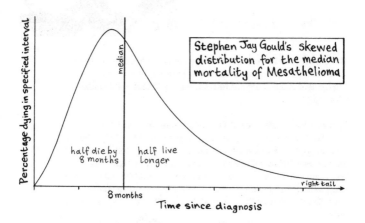

Stephen Jay Gould's skewed distribution for the median mortality of Mesathelioma

Percentage dying in specified interval

median

half die by 8 months

half live longer

right tail

8 months

Time since diagnosis

Gould reckoned that most people, without training in statistics, would have read "median mortality of eight months" to mean: "I will be dead in eight months",

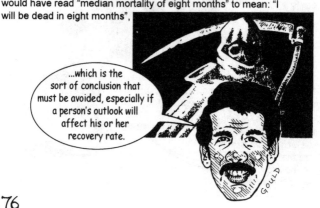

...which is the sort of conclusion that must be avoided, especially if a person's outlook will affect his or her recovery rate.

GOULD

As an evolutionary biologist, Gould learned to treat variation as a basic reality and to be wary of averages, which were, after all, abstract measures applicable to no single person and often largely irrelevant to individual cases.

Variation within whole systems is the ultimate reality, and the abstract nature of the average has a limited utility.

In light of his perspicacity, one *Sunday Times* columnist surmised rather astutely that "Statistics are a condemned man's best friend". Stephen Jay Gould died in 2002, fully two decades after the initial diagnosis.

Data Management Procedures

The Victorians were among the first to use statistics as a means to study mass phenomena. Colossal amounts of data were collected by state agencies, private organizations and various individuals interested in such social phenomena as poverty, disease and suicide. These are the main procedures they used to manage their data:

1. Tabulation – simply entering data in long columns of numbers.
2. Creating pie charts and various diagrams.
3. Reducing the data to create smaller subsets. For example, when Galton worked with a larger sample size, he often reduced the sample to 100 because of the explanatory power of percentages.

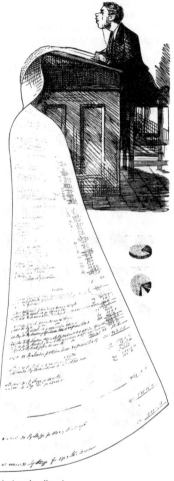

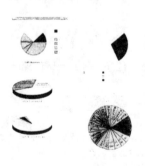

But since the charts or tables were not standardized, generalizations or comparisons with other data sets could not be made. Although they used averages to summarize this data, their statistical tools did little to convey its complexity, inherent in patterns of statistical variation.

Standardized Frequency Distributions

Pearson recognized that there had to be other ways to manage unwieldy data. He developed a systematic way to handle very large data sets by introducing the means to construct **standardized frequency distribution**. This allowed comparisons and generalizations about data sets that had previously been impossible to make.

The basic data-management procedures that Pearson introduced and the statistical methods he devised form the foundations of elementary mathematical statistics.

These are discussed in the following pages.

Samples vs. Populations

Pearson's closest friend, the Darwinian zoologist W.F.R. Weldon, began to use the term "sample" in 1892 to refer to collections of observations of marine organisms, though he wondered if his sample size was large enough. Pearson used the word "population" four years later to replace the term "normal group" and aligned *population* with *sample* in 1903.

Weldon

Karl Pearson

I advocate using very large samples because the results will be more representative of the entire population.

A *population* is a technical term for the whole group of organisms or objects, such as roses or tigers, for which the results are applicable. A population represents all conceivable observations of a particular type, whereas a *sample* is a limited number of observations from the population. The best example of using an entire population is the decennial Census count.

Population

Sample

In most studies, the population in which one is interested is far too large to measure each and every one of its members (all students in England, all voters in Britain, all cars made at Ford Motors, etc.). The statistician usually confines his or her analysis to a relatively small section of the total population.

Statisticians use different types of sampling techniques:
Random; **Systematic**; **Incidental**; **Purposive**; **Stratified**.

Random

This is analogous to putting everyone's name into a hat and drawing out several names. Everyone in the original population has an *equal* and *independent chance* of being included in the sample. Although this is the preferred way of sampling, it requires a complete list of every element in the population, which isn't always possible. A table of random numbers in statistics books or those generated by computers and some phone systems can be used.

Systematic

This also requires the entire listing of a population, but here it's divided into blocks where every nth person is selected from the list (e.g., by taking every 10th person from an alphabetized list).

Incidental

Uses the most accessible and available sample by taking the most convenient set of subjects. It is the most unreliable type of sampling.

Purposive

The experimenter chooses the subjects to be used because he or she thinks they're representative.

Stratified

The investigator selects a specific characteristic in the sample that he or she thinks is important for the research and then divides the sample into non-overlapping groups or strata, such as age-groups, gender, geographical areas or political affiliation. This can be used with the other four sampling procedures.

The Histogram

Pearson introduced the histogram on 18 November 1891. It is a term that he coined to designate a "time-diagram" in his lecture on "Maps and Chartograms".

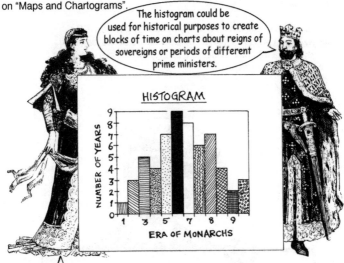

> The histogram could be used for historical purposes to create blocks of time on charts about reigns of sovereigns or periods of different prime ministers.

HISTOGRAM

NUMBER OF YEARS vs ERA OF MONARCHS

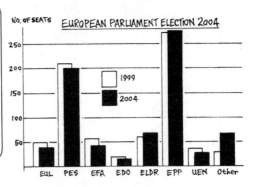

> The histogram is a graphical version of a set of continuous data (such as time, inches and temperature) that shows the number of cases that fall into adjacent rectangular columns that are contiguous (i.e. there are no gaps between the bars).

EUROPEAN PARLIAMENT ELECTION 2004

No. OF SEATS

1999
2004

EUL   PES   EFA   EDO   ELDR   EPP   UEN   Other

A similar-looking graph is a bar chart, but there are gaps between the bars and it uses discrete data (such as gender and political affiliation). Graphs are often used to help people think about a problem in visual terms.

A different way to depict the same set of continuous data is by using a frequency polygon. This is a line graph that joins the mid-points of each individual bar (from a histogram) together with a straight line.

This process of plotting data in the frequency polygon is the simplest type of "curve-fitting", which involves connecting two points of data together, either in a straight line or in a curve that produces various shapes.

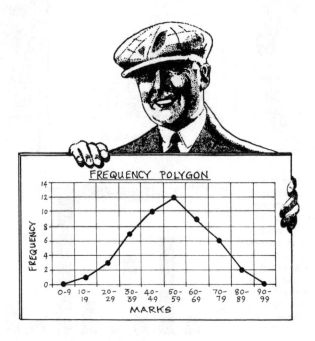

The next stage was for Pearson to show his students how to assemble frequency distributions for larger quantities of continuous data, and how to construct these distributions.

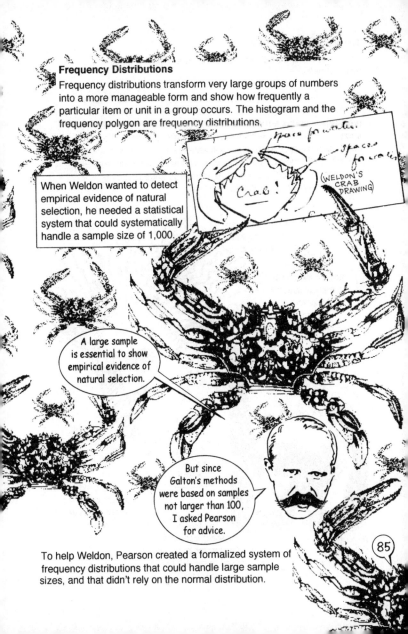

Frequency Distributions

Frequency distributions transform very large groups of numbers into a more manageable form and show how frequently a particular item or unit in a group occurs. The histogram and the frequency polygon are frequency distributions.

(WELDON'S CRAB DRAWING)

Crab!

When Weldon wanted to detect empirical evidence of natural selection, he needed a statistical system that could systematically handle a sample size of 1,000.

A large sample is essential to show empirical evidence of natural selection.

But since Galton's methods were based on samples not larger than 100, I asked Pearson for advice.

To help Weldon, Pearson created a formalized system of frequency distributions that could handle large sample sizes, and that didn't rely on the normal distribution.

The Method of Moments

How to determine and describe the shape of the empirical distribution?

Pearson began to develop his statistical system in 1892, based on the **method of moments**. The term "moment" originates in mechanics, and is a measure of force about a point of rotation, such as a fulcrum. In statistics, moments are averages; the computational procedures for moments are the same as finding the arithmetic mean. Pearson replaced mechanical force with a frequency curve function (such as the percentage of the distribution within a given class interval).

The first moment measures the average or the mean.

The second moment measures the average squared deviation …

… I termed this the "variance" in 1918.

The third moment measures the average cubed deviation (or skewness).

The fourth moment measures the average deviation raised to the 4th power (or kurtosis).

R.A. Fisher

An enthusiast for graphical representations, Pearson demonstrated the method of moments to his students by using examples from mechanics. To calculate the mean, he found the point about which a lever balances on a fulcrum. The mean is the "balance point" of this lever and is equivalent to the centre of gravity (or mass) in mechanics.

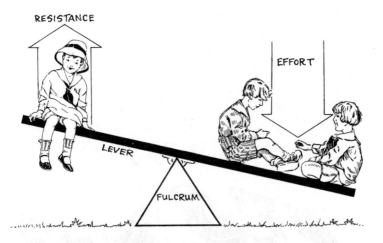

If a force is applied to the lever, then the first moment is called the "moment of force". Calculations are made to determine the first moment in order to find the mean. Pearson continued this procedure with the next three moments. Using the same data to find the mean, he squared these values to find the square of the *standard deviation* (see pages 99-102).

I called it the "squared standard deviation".

To find a measure of skewness of a distribution, he cubed these averaged values and calculated the third moment. When a distribution is skewed, the mean tends to lie closer to the tail.

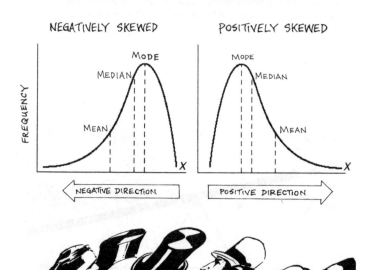

NEGATIVELY SKEWED

POSITIVELY SKEWED

MODE

MEDIAN

MEAN

FREQUENCY

NEGATIVE DIRECTION

MODE

MEDIAN

MEAN

POSITIVE DIRECTION

For skewness:
If the value = 0, this means the distribution is symmetrical.
A negative value = a negatively skewed distribution.
A positive value = a positively skewed distribution.

Pearson's first coefficient of skewness enabled him to calculate the asymmetry by subtracting the difference between the mean and the mode and dividing that by the standard deviation.

$$\text{Skewness} = \frac{\text{mean} - \text{mode}}{\text{standard deviation}}$$

To find the fourth moment he raised these averaged values to the fourth power. This measures how flat or peaked is the curve of the distribution. Pearson coined the word *kurtosis* for it (from the Greek word for bulginess), and it has three further components.

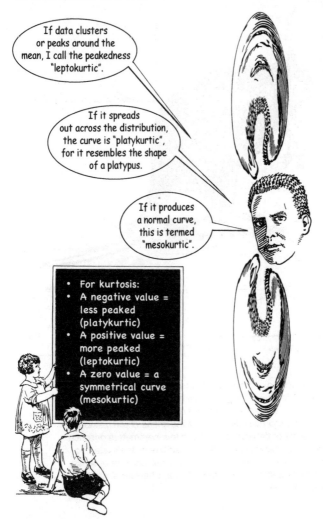

If data clusters or peaks around the mean, I call the peakedness "leptokurtic".

If it spreads out across the distribution, the curve is "platykurtic", for it resembles the shape of a platypus.

If it produces a normal curve, this is termed "mesokurtic".

- For kurtosis:
- A negative value = less peaked (platykurtic)
- A positive value = more peaked (leptokurtic)
- A zero value = a symmetrical curve (mesokurtic)

One of Pearson's students, **William Sealy Gosset** (1876–1937), who adopted the pseudonym "Student", used an illustration of a platypus to show a platykurtic curve and two long-tailed kangaroos for a leptokurtic curve.

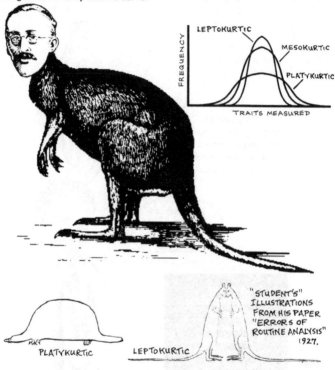

"STUDENT'S" ILLUSTRATIONS FROM HIS PAPER "ERRORS OF ROUTINE ANALYSIS" 1927.

From the method of moments, Pearson established four parameters for curve-fitting to show how the data *clustered* (the mean), how it *spread* (the standard deviation), if there were a *loss of symmetry* (skewness) and if the shape of the distribution were *peaked or flat* (kurtosis). These four parameters describe the essential characteristics of any distribution: the system is parsimonious and elegant. These statistical tools are essential for interpreting any set of statistical data, whatever shape the distribution takes.

Natural Selection: the Changing Shapes of Darwinian Distributions

Darwin understood that the shape of a frequency distribution before natural selection occurred would be "symmetrical about the mean" (i.e. normally distributed) and that with selection, the distribution would lose its symmetrical bell-shaped curve. (However, following reproduction, the normal curve will be restored, but with a different mean value.)

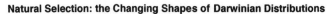

Natural selection, the mechanism of evolution, is caused by variation in Darwinian fitness (i.e. suitability to the environment). It is genetically determined and measured by differential fertility and/or mortality rates.

Thus, only the organisms that are best adapted to their environment tend to survive and transmit their genetic characteristics to succeeding generations, while those less adapted tend to be eliminated.

If the shape of the distribution is peaked or flattened (kurtotic, to use Pearson's term), this would suggest *stabilizing selection*, which seeks to maintain the status quo.

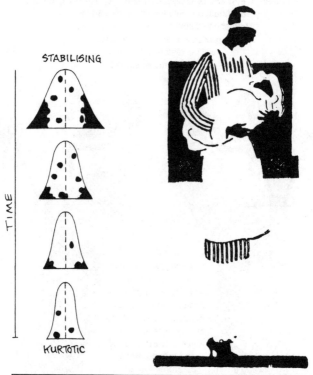

The distribution at the top is normally distributed before selection: the areas in black indicate where selective pressure happens over a period of time, until the shape of the distribution eventually changes to the one at the bottom.
Selective pressure is any phenomenon which alters the behaviour and fitness of living organisms within a given environment. It is the driving force of evolution and natural selection.

The birth weight of human infants is under a stabilizing selection. Infant mortality is lowest at intermediate birth weights and highest at low and high birth weights.

A distribution that is bimodal would mean *disruptive selection,* which weeds out the middle of the distribution and favours those on both ends of it. Disruptive selection is found in the black-bellied seedcracker (*Pyrenestes ostrinus*) which lives in West Africa. Birds with small beaks eat soft, small seeds, while birds with large beaks eat large, hard seeds.

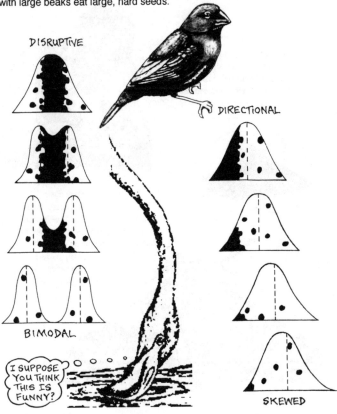

DISRUPTIVE

DIRECTIONAL

BIMODAL

SKEWED

I SUPPOSE YOU THINK THIS IS FUNNY?

If a distribution becomes skewed in one direction, this would indicate *directional selection,* which happens when a population finds circumstances more favourable at one end of the distribution than the other.

The Peppered Moth

One of the best-known examples of directional selection is the peppered moth (*Biston betularia*), which was abundant in pre-industrial Victorian England. Though a coal-black mutant was found in 1849, this was a rare occurrence at the time.

In highly industrialized Victorian cities like Manchester and Leeds, air pollution was severe and toxic gases and soot blackened tree trunks.

The black moth, practically invisible on the darkened tree trunks, had a better chance of survival than the peppered ones, which were easily picked off by birds.

(see page 127)

Within a century these coal-black moths had increased to 90% of the population in the industrial north. While the original peppered moths formed a normal distribution, once the area where the moths lived became polluted, the normal curve shifted to the right side of the distribution, producing a skewed curve.

The Pearsonian Family of Curves

By calculating the method of moments, Pearson also provided a variety of theoretical curves in varying graduations, which could then be superimposed onto an empirical curve to determine which gave the best "fit". These curves were referred to as the "Pearsonian family of curves".

Gamma Curve

t-distrib-ution

The more important ones that remain an essential part of theoretical statistics today include:

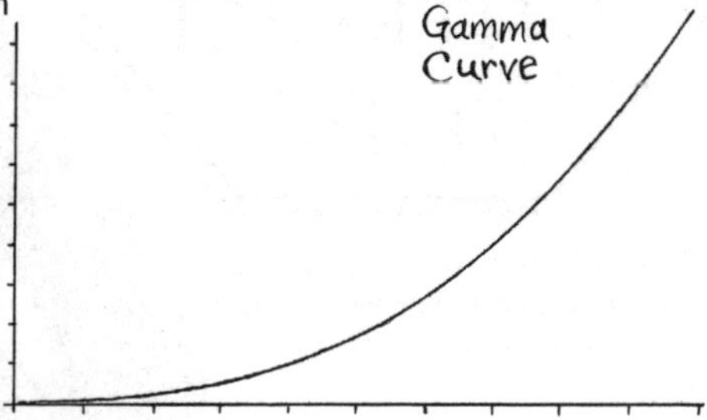

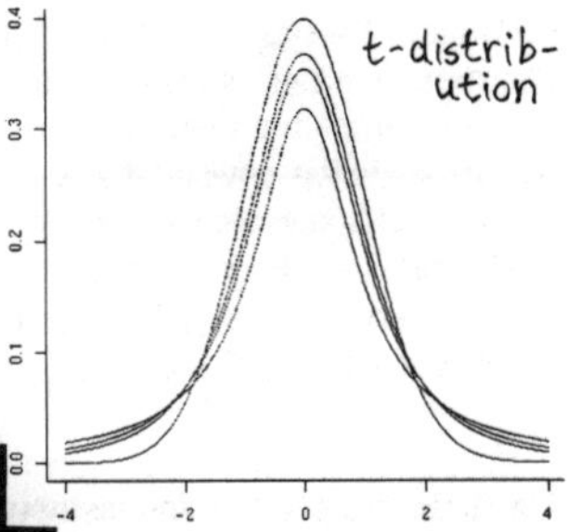

Type III, the Gamma Curve, which he went on to use for finding the exact chi-square distribution (discussed later)
Type IV, the Family of Asymmetric Curves (created for Weldon's data)
Type V, the Normal Curve
Type VII, now known as Student's distribution for t-tests (examined later)

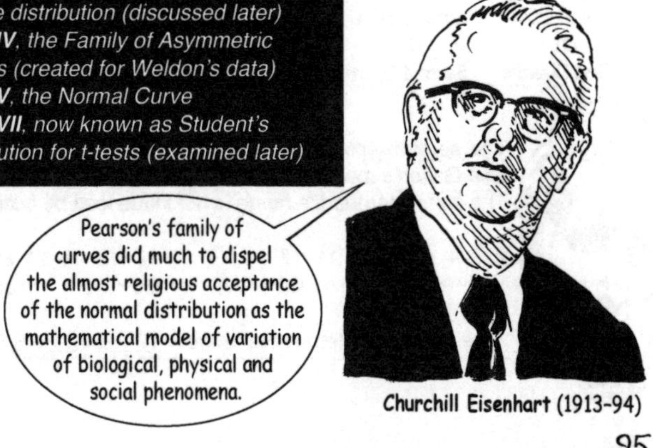

Pearson's family of curves did much to dispel the almost religious acceptance of the normal distribution as the mathematical model of variation of biological, physical and social phenomena.

Churchill Eisenhart (1913–94)

HOW TO INTERPRET DATA?

The statistician begins by looking for overall patterns of variation and any striking deviation from that pattern.

Statistical Measures of Variation

The measurement of variation is the lynchpin of mathematical statistics. Galton devised the first measure of statistical variation in 1875 when he introduced the "semi-interquartile range", which he expressed as:

$$\frac{Q3 - Q1}{2}$$

A quartile is a point on the distribution.

1% to 25%	26% to 50%	51% to 75%	76% to 100%
Q1	**Q2**	**Q3**	**Q4**
First Quartile	**Second Quartile**	**Third Quartile**	**Fourth Quartile**

The semi-interquartile range

Like Galton's median, this method was quick and easy to use. The semi-interquartile range is not influenced by outliers:

2 3 4 6 **6** 8 9 11 **12** 14 14 15 **17** 18 19 21 **82**
       Q1   **Q2**   **Q3**   **outlier**

Here, the semi-interquartile range = $\frac{17 - 6}{2} = \frac{11}{2} = 5.5$

GALTON

96

The Interquartile Range

This is a more widely-used method that measures the spread of the middle 50% (or the median) of an ordered data set. In the example:

1 1 3 4 **4** 5 5 6 **6** 7 7 8 **8** 9 9 9 10
     Q1   **Q2**   **Q3**

... the interquartile range = Q3 – Q1 or 8 – 4 = 4. Thus, the median (Q2) equals 6 and spreads across a 4-point range. This technique remained a quick and easy way to hand-calculate an approximate estimate of variation, until the advent of statistical software packages for PCs in the late 1970s.

Like the semi-interquartile range, the interquartile range is not influenced by outliers:

2 3 4 6 **6** 8 9 11 **12** 14 14 15 **17** 18 19 21 **82**
    Q1   **Q2**   **Q3**   **outlier**

The interquartile range = Q3 – Q1 or 17 – 6 = 11; the median of 12 spreads across an 11-point range.

THE RANGE

In his first Gresham Lectures on statistics in 1892, Pearson introduced the **range**, which is the simplest method used to measure variation. The range measures the distance between the largest and smallest values from a particular set of measurements and gives an idea of the spread of the data.

The range in this example: 4, 7, 12, 25, 34 would be 34 – 4 = 30.

It's quite often used for summaries of data made available to the general public, such as the range of salaries, ages and temperatures.

The virtue of the range is its simplicity, but it's the least reliable measure of variation, as it doesn't use all data and is also affected by outliers.

In this example of Celsius temperatures in one week in November: 2, 6, 8, 12, 10, 12, 26, the range = 26 – 2 = 24.

The result of 24°C is not a reliable numerical gauge of the full range of temperatures in a week in November. The unseasonable high of 26°C is an anomaly (or, perhaps, an indicator of global warming).

98

The Standard Deviation

Pearson introduced the **standard deviation** in his Gresham lecture of 31 January 1893, referring to it initially as the "standard divergence". John Venn had used the term "divergence" a couple of years earlier when referring to deviation. The standard deviation is a measure of variation. It indicates how widely or closely spread the values are in a set of a data, and shows how much each of these individual values deviate from the average (i.e. the mean).

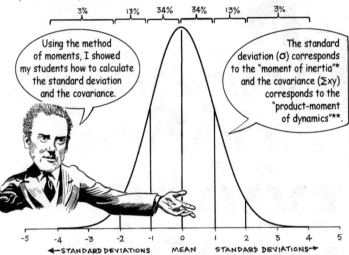

The covariance is the measure of how much two random variables move together. If two variables tend to move together in the same direction, then the covariance between the two variables will be positive. If two variables move in the opposite direction, the covariance will be negative. If there is no tendency for two variables to move one way or the other, then the covariance will be zero.

---

*The moment of inertia is an important value in mechanics: it is a geometrical property of a beam, and a measure of the beam's ability to resist buckling or bending.

**The moment of dynamics is concerned with the effect of force on the motion of objects.

By using the standard deviation, Pearson made it possible to measure *all* the points of variation on a distribution rather than the two or three points that Galton had offered in his quartile range.

> The standard deviation shows the deviation from the mean and the frequency of this deviation.

> It has remained, without doubt, one of the most widely used statistical tools to measure variation.

The standard deviation = $\sqrt{\dfrac{(\text{sum of raw scores} - \text{mean of observations})^2}{\text{number of observations}}}$

or

$$S = \sqrt{\frac{\Sigma(X - \overline{X})^2}{N}}$$

Thus, the standard deviation = the square root of the average squared deviations.

Instead of simply adding up values to find the mean, here we...

(1) subtract the mean from the raw scores (X), which provides the "deviational" value (symbolized by little x);
(2) disregard the positive and negative values by squaring this set of values;
(3) add up the deviational squared values to calculate the average squared deviation (or the standard deviation).

Raw score	–	mean	=	Deviational value	Deviational value squared
X		**x̄**		**x**	**x^2**
12		8		4	16
10		8		2	4
6		8		-2	4
8		8		0	0
<u>4</u>		8		-4	<u>16</u>
<u>40</u>					40
5 = 8					

Deviational formula:

$$S = \sqrt{\frac{\sum x^2}{n}} = \sqrt{\frac{40}{5}} = \sqrt{8} = 2.82$$

This means that the average amount of deviation in this set of data is 2.82 units away from the mean value of 8 and that, therefore, there is a small amount of variation in this sample.

The standard deviation is expressed in the same units as the raw data.

That is, if something is measured in feet, inches or centimetres,...

...the standard deviation is expressed in feet, inches or centimetres.

A large standard deviation (relative to the value of the mean) shows that the frequency distribution is widely spread out from the mean, whereas a small standard deviation indicates that it lies closely concentrated near the mean, with little variability between one observation and another. Although the standard deviation indicates to what extent the whole group deviates from the mean, it does not show how variable a particular group is.

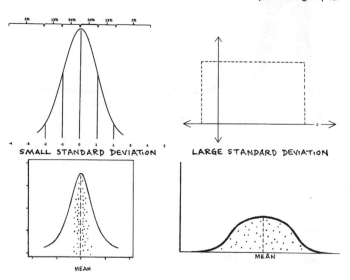

SMALL STANDARD DEVIATION

LARGE STANDARD DEVIATION

MEAN

MEAN

While the standard deviation is a practical measure of variation, the variance is used for theoretical work, especially with the *analysis of variance* (see pages 168-70).

The variance is also a measure of variation, but it is used for random variables and indicates the extent to which its values are spread around the expected values.*

Using the example from the standard deviation:

The variance =
$$\frac{(\text{sum of squared deviations} - \text{mean of observations})^2}{\text{number of observations}}$$

or

$$S^2 = \frac{\sum(X - \bar{x})^2}{N}$$

The deviational formula for the variance is:

$$S^2 = \frac{\sum x^2}{N} = \frac{40}{5} = 8$$

* The expected values represent the average amount one "expects" as the outcome of the random trial when identical odds are repeated many times.

103

Since the standard deviation doesn't show the range of variation within a group, how did Pearson determine how variable a particular group was and how to make comparisons with other groups with widely different means? For this he needed a different statistical method.

This was a problem that I encountered in 1886 when I was measuring the heights of men and women.

INVISIBLE "LIFTEE" HEIGHT **PAD**

I wanted to find out whether it was the men or the women who were the more variable in height.

Galton coped with this problem by adjusting the mean stature of women with an equivalent mean stature of men and then comparing the deviations in men and women. He adjusted or "transmuted" all female statures to male by multiplying them by the constant 1.08.

Coefficient of Variation

Pearson thought the best way to compare deviations in heights of men and women was to alter the deviations in the same ratio. The use of the standard deviation alone, which could be measured in centimetres or inches, would most likely show that men are taller on average, since they would have a higher mean value, but this wouldn't answer the question:

"Who shows more variation within their group?"

Pearson devised the **coefficient of variation** to measure this. This was important to Pearson when he was trying to determine how variable some of Weldon's prawns and crabs were.

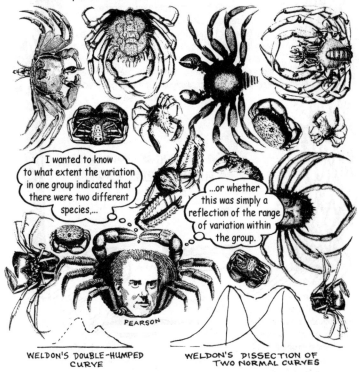

I wanted to know to what extent the variation in one group indicated that there were two different species,...

...or whether this was simply a reflection of the range of variation within the group.

WELDON'S DOUBLE-HUMPED CURVE

WELDON'S DISSECTION OF TWO NORMAL CURVES

105

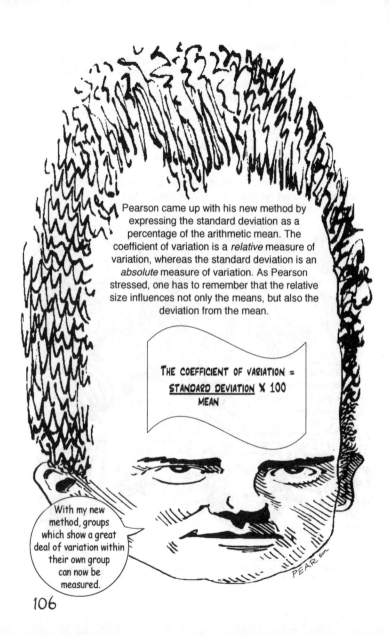

Pearson came up with his new method by expressing the standard deviation as a percentage of the arithmetic mean. The coefficient of variation is a *relative* measure of variation, whereas the standard deviation is an *absolute* measure of variation. As Pearson stressed, one has to remember that the relative size influences not only the means, but also the deviation from the mean.

THE COEFFICIENT OF VARIATION =
$$\frac{\text{STANDARD DEVIATION}}{\text{MEAN}} \times 100$$

With my new method, groups which show a great deal of variation within their own group can now be measured.

Comparing Variation of Variables

The coefficient of variation has no units, so one can use it to compare variation for different variables with different units. Thus, comparisons can be made for Celsius in London and Fahrenheit in New York in one week to determine which set of temperatures is more variable.

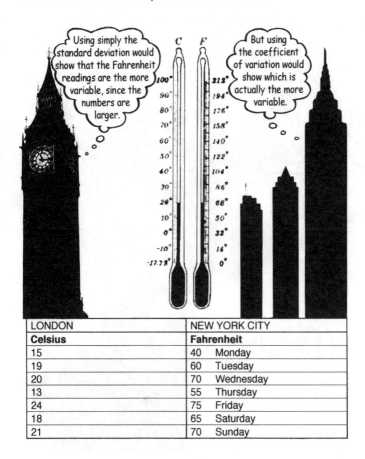

LONDON	NEW YORK CITY	
Celsius	**Fahrenheit**	
15	40	Monday
19	60	Tuesday
20	70	Wednesday
13	55	Thursday
24	75	Friday
18	65	Saturday
21	70	Sunday

Practical Applications

This method remains a very widely used tool for industry, marketing and economics. Wool manufacturers use the coefficient of variation to calculate the variation in fibre diameter distribution and yarn irregularities.

This information, for example, enables manufacturers to produce various qualities of wool, depending on the demands of the market.

Pearson's Scales of Measurement

Distinguishing scales of measurement was crucial in the development of Pearson's methods of correlation and for other statistical tests. When Galton, Weldon and Pearson first began to analyse data, virtually all of it could be described as "continuous" data. By 1899, Pearson had begun to work on statistical coefficients to measure relationships between "discontinuous" (or discrete) variables.

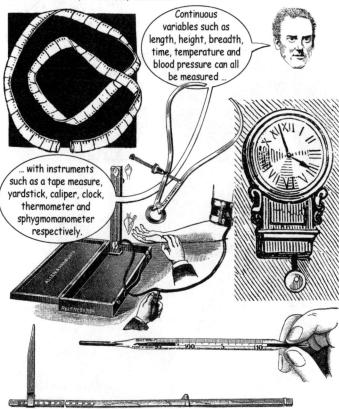

Continuous variables such as length, height, breadth, time, temperature and blood pressure can all be measured ...

... with instruments such as a tape measure, yardstick, caliper, clock, thermometer and sphygmomanometer respectively.

These variables are expressed in units of measurements that can be broken down into definite gradations such as inches, centimetres, seconds, minutes or degrees.

Nominal and Ordinal Variables

Pearson first encountered variables that could not be treated as continuous when he began to look at the inheritance of eye colour in man and colour of coat in horses and dogs. In these situations, the only form of classification available for the variables is one that involves "counting", rather than "measuring": eye colour cannot be measured in the same way that stature, weight or time can be measured.

Pearson referred to such variables as eye colour as **nominal**.

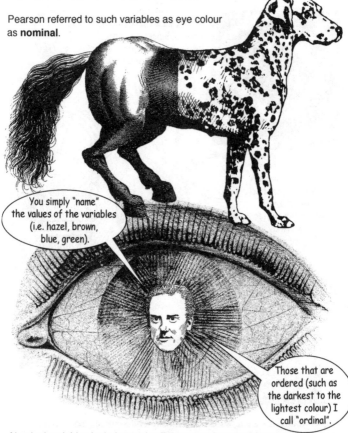

You simply "name" the values of the variables (i.e. hazel, brown, blue, green).

Those that are ordered (such as the darkest to the lightest colour) I call "ordinal".

Nominal variables include nearly all demographic variables such as religious affiliation, political persuasion and socio-economic status.

Ordinal variables are simply ordered and then named. The Mohs Scale, devised by the German mineralogist Friedrich Mohs in 1822, is an example of an ordinal scale.

Hardness	Mineral	Absolute Hardness
1	Talc	1
2	Gypsum	3
3	Calcite	9
4	Flourite	21
5	Apatite	48
6	Orthoclase Feldspar	72
7	Quartz	100
8	Topaz	200
9	Corcumdum	400
10	Diamond	1500

It consists of ten minerals, from the hardest (diamond) to the softest (talc), and measures their hardness or scratch resistance.

It doesn't imply equal degrees of difference in hardness: a diamond is not 100 times harder than talc, it's simply the hardest of all minerals.

For statistical results to be valid and meaningful, it's essential to use the right kinds of statistical methods for the type of data one is studying.

111

Ratio and Interval

The American psychologist **Stanley Smith Stevens** (1906–73) made a further sub-division with "continuous variables" in 1947 when he introduced *ratio* and *interval* scales of measurement (most of Pearson's continuous variables were ratio). Stevens proposed the following:

1. Ratio scales

These differ from interval variables (see next page) in two ways: a) an absolute zero indicates the absence of the property being measured (i.e. height, weight and blood pressure) and b) ratio scales are additive.

Thus, one can say that something is "twice as tall" or "three times the distance".

As metric and imperial tools are absolute measures, the difference between 3 feet and 6 feet is identical to a metric reading of the difference between 0.91 metres and 1.81 metres.

Both are twice as long.

S.S.S.

2. Interval scales

The zero point is arbitrary and does not reflect the absence of an attribute (such as 0 Celsius and 0 Fahrenheit readings).

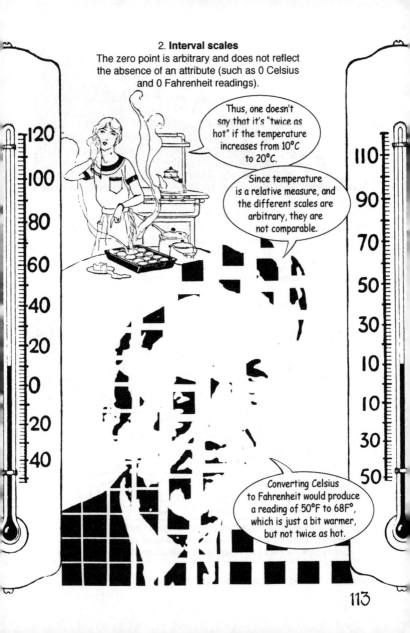

Thus, one doesn't say that it's "twice as hot" if the temperature increases from 10°C to 20°C.

Since temperature is a relative measure, and the different scales are arbitrary, they are not comparable.

Converting Celsius to Fahrenheit would produce a reading of 50°F to 68F°, which is just a bit warmer, but not twice as hot.

Correlation,

one of the most widely used statistical methods, indicates the extent to which two variables go together (e.g., height and weight). The most common type measures a linear relationship between two variables, and refers to how well they go together in a straight line.

But not every pair of characters or variables can be assessed by using a statistical correlation, and different methods of correlation are used within the biological, medical, behavioural, social and environmental sciences, as well as in industry, commerce, economics and education.

Different types of correlational methods are used for different types of variables, depending on the scale of measurement.

Variables can be nominal, ordinal, interval or ratio.

1950 · 1970 · 2010

And data can be ranked and can take on categories such as binary (0,1) and dichotomous (two mutually exclusive values), which require the use of very specific methods of correlation.

Pearson devised methods for all types of variables.

Early Uses of Correlation

The word "correlation" had already been in use a century before there was a way to measure it. The word was first used by the biologist **Comte de Buffon** (1707–88) and was further developed by the palaeontologist **Baron Georges Cuvier** (1769–1832) who talked about the "correlation of parts" in 1801.

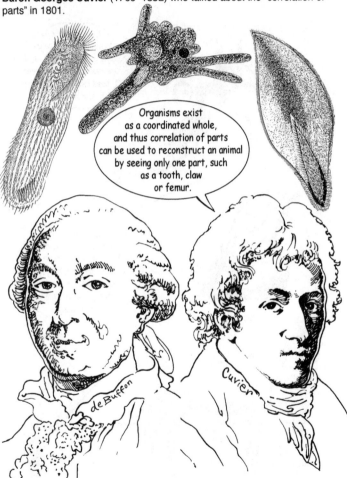

Organisms exist as a coordinated whole, and thus correlation of parts can be used to reconstruct an animal by seeing only one part, such as a tooth, claw or femur.

de Buffon

Cuvier

Charles Darwin, who thought Cuvier's idea of correlation was important, discussed *functional correlations* when, for example, the size of one organ is a function of a second organ. He also discussed *developmental correlation*, which arose in the early stages of growth and influenced the development of an organism.

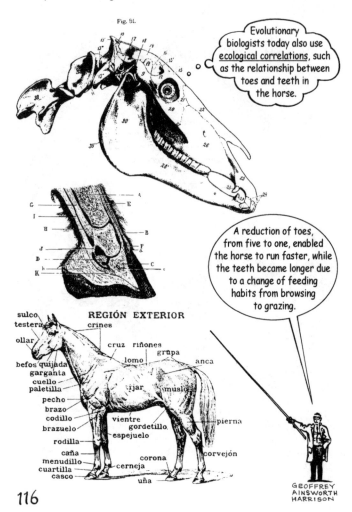

Fig. 51.

Evolutionary biologists today also use <u>ecological correlations</u>, such as the relationship between toes and teeth in the horse.

A reduction of toes, from five to one, enabled the horse to run faster, while the teeth became longer due to a change of feeding habits from browsing to grazing.

REGIÓN EXTERIOR

sulco
testera
crines
ollar
cruz riñones
grupa
lomo anca
befos quijada
garganta
cuello
paletilla
pecho
brazo
codillo
brazuelo
rodilla
caña
menudillo
cuartilla
casco
cerneja
corona
uña
vientre
gordetillo
espejuelo
ijar muslo
pierna
corvejón

GEOFFREY
AINSWORTH
HARRISON

Causation and Spurious Correlation

Francis Galton was the first person to come up with a method to measure correlation when he created a graph to find a relationship between mother and daughter sweet peas.

Until Galton invented the idea of correlation, **causation** was the primary way in which two related events were explained, especially in the physical sciences.

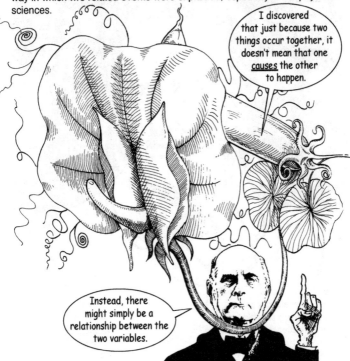

I discovered that just because two things occur together, it doesn't mean that one <u>causes</u> the other to happen.

Instead, there might simply be a relationship between the two variables.

Before meeting Galton, Pearson was convinced that formal mathematics could be applied only to natural phenomena that were determined by causation. But Galton's ideas of correlation replaced causation for Pearson, especially in the biological sciences. He became an anti-causalist who thought that the universe was not controlled by laws of causation, in its narrowest sense, but that variation instead played a bigger role in helping to explain phenomena.

Pearson warned his students that correlation should not be interpreted as an indication of causation, though he realized that "to those who persist in looking upon all correlation as cause and effect, the fact that correlations can be produced between two uncorrelated characters, must come as rather a shock". Moreover, the direction of the cause is unknown: Does X cause Y or does Y cause X?

Not all correlations are genuine, and it's possible to find a mathematically perfect correlation that is completely meaningless.

These I call <u>spurious correlations</u>.

Hence, a mathematically perfect correlation does not mean causation: it simply means that two variables are very highly correlated. This may even be the result of a spurious or illusory correlation due to the influence of a third variable, called a "lurking variable". While students' university qualifications are highly correlated with their income later in life (the higher the grades, the higher the salary), this correlation could be due to a third (lurking or hidden) variable, such as the tendency to work hard.

Path Analysis and Causation

The evolutionary biologist Sewall Wright extended Pearson's ideas of correlation into the realm of cause and effect by mapping out the logical and methodological relationships between correlation and causation. Using Pearson's multiple regression (see pages 134-8), Wright devised a statistical methodology in 1918 that he called *path analysis*.

Scatter Diagrams

Correlation is often depicted graphically on something called a scatter diagram to see what shape it produces. If two variables produce a narrow ellipse that resembles a straight line, this would indicate a high correlation. A full-size ellipse reveals a moderate correlation, whereas a circle indicates no correlation. In this way, correlation measures the *strength* (high, medium or low) of the relationship.

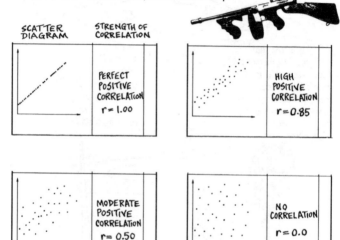

Correlation cannot, however, be transformed to a percentage. Thus, a moderate correlation of 0.55 or a high correlation of 0.80 is not equivalent to 55% or 80%, as some people erroneously believe.

Weldon and Negative Correlation

The numerical index that correlation yields also measures the *direction* of the relationship. Either two variables move up or down the graph together (e.g., height and weight in healthy infants goes up together) or one variable moves up while the other moves down (e.g., the faster one travels in a car, the sooner the destination is reached: speed increases as time decreases). The former produces a positive or direct correlation while the latter yields a negative or inverse correlation.

I suggested to Pearson in 1896 the idea of a "negative" or "inverse" correlation.

Hence, the values of the correlation coefficient could range from –1.00 to +1.00 rather than the range 0.00 to +1.00, as suggested firstly by Galton.

W.F.R. Weldon with his wife & work colleague, Florence

SCATTER DIAGRAM | DIRECTION OF CORRELATION

PERFECT NEGATIVE CORRELATION

HIGH NEGATIVE CORRELATION
r = –.80

Curvilinear Relationships

Though the numerical index provides some information about the degree of a linear relationship, a scatter plot is a useful tool, because it may reveal instead a curvilinear relationship. Pearson introduced the correlation ratio in 1905 to measure a curvilinear relationship.

An age over lifetime growth curve is curvilinear, whereas an early childhood growth curve is linear. Infants continue to grow until adolescence: they get taller, their hair grows, and they also become more dextrous, agile and flexible. But the lifespan is curvilinear because a number of these characteristics diminish among the elderly: many become shorter, men in particular have a tendency to lose their hair and some become bald, and many others become less agile and flexible as they get older.

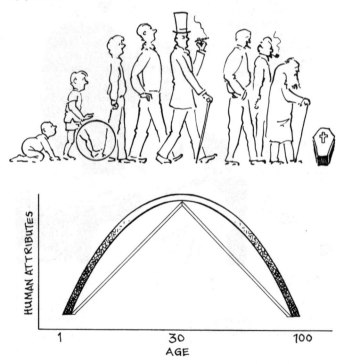

Galton and Biological Regression

Before Galton's work on correlation, he was writing about *regression*.

I wanted to know how it was possible for successive generations to remain alike in so many features ...

... and also why offspring vary – some will be taller and others shorter than their parents.

Galton measured the diameter and weight of thousands of mother and daughter sweet pea seeds in 1875, and found that the population of the offspring reverted towards the parents and followed the normal distribution. As the size of the mother pea seed increased, so did the size of the daughter pea seed, but the offspring was not as big or as small as the mother pea; it therefore regressed back towards the size of its "ancestor pea".

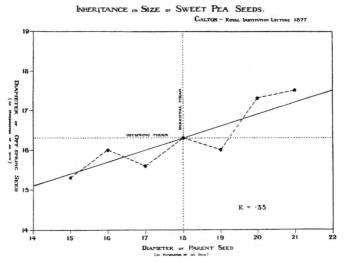

GALTON'S SWEET PEA REGRESSION LINE

Regression to the Mean

This refers to the tendency of a characteristic in a population to move away from the extreme value and closer to the average values.

Galton wanted to find the correlation of height between fathers and sons, because it was easy to measure and it remained stable during adult life.

I realized that correlation was bi-directional and produced two regression lines: one for the offspring on parents and the other for the parents on offspring.

However, Galton's realization created a paradox, for it contradicted what he understood to be the uni-directionality of regression. Galton thus had to explain how the height of the offspring could in any way influence the height of the parents.

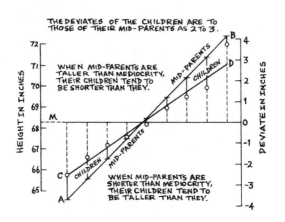

Galton's Two Regression Lines

While Galton showed that there was a correlation between fathers and sons, his two regression lines gave a different picture. The lines in the upper part of the graph revealed that if parents were taller than average, then their children would be shorter than their parents – the children's value regresses to the mean. Conversely, the regression lines in the lower part of the graph indicated that, if the parents were shorter than average, their children would be taller than their parents, while regressing to the mean value.

The heights of fathers and sons is used to illustrate an individual case of a regression to the mean:

> ### Table A
>
> **Regression of father's height on son's height**
>
> Father = 6'1"
> Mean = 5'6"   ('= feet
> Son = 5'9"   ("= inches

> ### Table B
>
> **Regression of son's height on father's height**
>
> Son = 6'2"
> Mean = 5'8"
> Father = 5'6"

In the first table, the mean height taken from a sample of 100 fathers and their sons is 5'6" and the height of a father is 6'1". If we then regress the height of sons to fathers, the son's height is 5'9". The father is taller than the average height but the son is shorter than the father; thus, the value has regressed to the mean. In the second table, where the son's height is 6'2", a regression of the father's height in this example could lead to a height of 5'8" for the father. Here the father is shorter than average, but his son is taller than his father.

ROBERT PERSHING WADLOW (1918-1940) – TALLEST MAN IN THE WORLD WITH 'NORMAL' ADULT & CHILD.

Since regression to the mean refers to the tendency of a characteristic in a population to move away from extreme values and closer towards average values, this reinforced Galton's view that distributions would always be normal. He was convinced that natural selection could not produce permanent change in a population, as the next generation would regress towards the mean value for the species.

Galton did not take into account the fact that, following reproduction after natural selection had changed the shape of the distribution, the normal curve would indeed be restored, but with a different mean value. (See page 93.)

Regression, however, has no effect on the variation (or variance) of the population: the variation does not diminish because of the phenomenon of regression.

AVERAGE POPULATION SHIFTS IN ONE DIRECTION.

SOLID CURVE: ORIGINAL CURVE BEFORE SELECTION.
DOTTED CURVE: NEW NORMAL CURVE WITH A DIFFERENT MEAN VALUE AFTER SELECTION.

SOME REGRESSION!

MARSH-MALLOWS

George Udny Yule and the Method of Least Squares

At the end of the 19th century, Pearson's student **George Udny Yule** (1871–1951) introduced a novel approach to interpreting correlation and regression with a conceptually new use of the *method of least squares*, which is a mathematical tool that reduces the influence of errors when fitting a regression line to a set of data points.

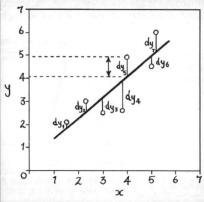

This method calculates the best-fitting line for the observed data by minimizing the sum of the squares of the vertical deviations from each data point to the regression line.

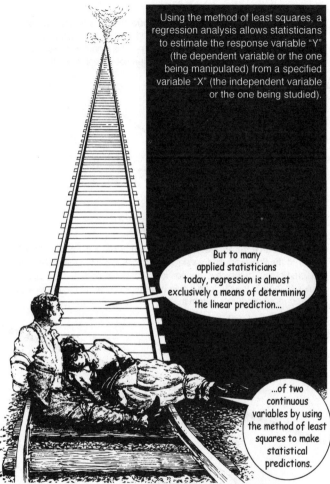

Using the method of least squares, a regression analysis allows statisticians to estimate the response variable "Y" (the dependent variable or the one being manipulated) from a specified variable "X" (the independent variable or the one being studied).

But to many applied statisticians today, regression is almost exclusively a means of determining the linear prediction...

...of two continuous variables by using the method of least squares to make statistical predictions.

Although the method of least squares may be used to analyse regression lines, much of the confusion surrounding regression to the mean can be attributed to those who forget that Galton's regression to the mean involves *two* regression lines and not simply one regression line to be used to predict future outcomes by using the method of least squares.

Correlation vs. Regression

Though Galton wanted to measure the correlation of stature between father and son, Pearson discovered in 1896 that Galton's procedure for finding "co-relation", as he spelt it, measured the *slope* of the regression line, which was a measure of the regression coefficient instead.

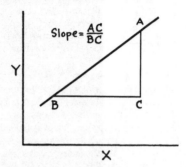

$$Slope = \frac{AC}{BC}$$

Galton fitted an arbitrary line and then tested to see if the slope of the line were 1. If the line produced a value of 1, this would have meant that the predicted height of the children was identical to the parent; if the value were less than 1, the children would have been closer to the mean and thus had more moderate heights.

Galton's Dilemma

How was it that when Galton was trying to produce a mathematical formula for correlation, he ended up with a measure of regression instead? Pearson clarified Galton's work.

I showed that Galton's mistake was to assume that there was "equal variability" between parent and offspring (i.e. that the standard variation should have produced identical numerical values).

Pearson was able to measure this variation of father and son separately by using his standard deviation. He then showed that if the standard deviations of a character in the offspring and the parent happen to have the same numerical values, then the regression coefficient and the correlation coefficient would indeed also have identical values. He stressed, however, that the correlation coefficient and the regression coefficient will almost always yield different values.

Hence, Galton conflated the concepts of correlation and regression in his work. Pearson had broken the uni-directionality of Galton's concept of regression, thereby freeing it from the narrow context of human heredity and transforming it into a purely statistical concept. Since Pearson had shown that Galton's correlation formula was instead a measure of regression, he retained Galton's "r" to symbolize the correlation coefficient.

Pearson's Product-Moment Correlation

Building on the system of the method of moments, Pearson devised a mathematically rigorous formula for correlation. He demonstrated that the optimum values of the regression slope and correlation coefficient could be calculated from the product moment where x and y are deviations of observed values from their respective means. Pearson found that the best formula for what he termed in 1896 the *product-moment correlation coefficient* was:

$$r = \frac{\Sigma(xy)}{(S_x)(S_y)} = \frac{\text{covariance}}{(\text{standard deviation of } x)(\text{standard deviation of } y)}$$

The covariance, $\Sigma(xy)$ is a measure of how much the deviations of two random variables move together. (See page 99.)

Pearson then determined that the regression coefficient was:

$$b = \frac{\Sigma(xy)}{S^2_x} = \frac{\text{covariance}}{\text{variance of } x}$$

PRODUCT-MOMENT CORRELATION COEFFICIENT

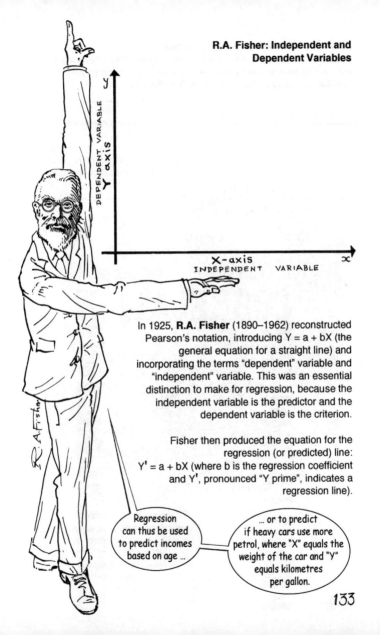

DEPENDENT VARIABLE Y axis

y

X-axis INDEPENDENT VARIABLE

x

R.A.Fisher

In 1925, **R.A. Fisher** (1890–1962) reconstructed Pearson's notation, introducing $Y = a + bX$ (the general equation for a straight line) and incorporating the terms "dependent" variable and "independent" variable. This was an essential distinction to make for regression, because the independent variable is the predictor and the dependent variable is the criterion.

Fisher then produced the equation for the regression (or predicted) line: $Y' = a + bX$ (where b is the regression coefficient and Y', pronounced "Y prime", indicates a regression line).

Regression can thus be used to predict incomes based on age ...

... or to predict if heavy cars use more petrol, where "X" equals the weight of the car and "Y" equals kilometres per gallon.

133

Simple Correlation and Multiple Correlation

Pearson introduced the term **simple correlation** when measuring a linear relationship between *two* continuous variables only, such as the relationship between stature of father and stature of son.

When I became interested in measuring the relationships between characters in <u>more than two</u> generations, I needed a different set of statistical procedures.

Francis Ysidro Edgeworth

Galton

I had already dealt with statistical correlations of three variables in 1892, which I expressed as "Galton's functions", named by Weldon in 1889.

When I offered a mathematical resolution for Galton's work, I devised the mathematical structure for multiple correlation, symbolized by R,

… to measure the relationship of three or more continuous variables (i.e. between one dependent variable and the combined set of two or more independent variables). Hence, multiple correlation involves the simultaneous calculation of the correlation coefficients of several variables.

This work provided the basis for the development of **multiple regression**. Like simple regression, it involves a linear prediction, but rather than using only one variable to be "predicted", a collection of variables can be used instead.

THE OVERLAP BETWEEN THREE VARIABLES

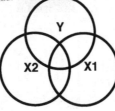

X1 = FOOD CONSUMPTION
X2 = AMOUNT OF EXERCISE
Y = BODY MASS INDEX

HIGHER MATHEMATICS AND MATRIX ALGEBRA

To calculate the multiple correlation coefficient, Pearson introduced a higher form of mathematics. This played a pivotal role in the professionalization of mathematical statistics as an academic discipline at the end of the 19th century. Pearson learnt this type of mathematics at Cambridge from J.J. Sylvester and **Arthur Cayley** (1821–95), who had created matrix algebra out of their discovery of the theory of invariants during the mid-19th century.

It wasn't until the 1930s that Pearson's higher form of mathematics was replaced by matrix algebra in mathematical statistics.

Matrix algebra has remained central to multivariate statistics.

Cayley

Sylvester

EXAMPLE OF MATRIX ALGEBRA

Matrix A
$$\begin{bmatrix} 7 & 3 \\ 2 & 5 \\ 6 & 8 \\ 9 & 0 \end{bmatrix}$$

Matrix B
$$\begin{bmatrix} 7 & 4 & 9 \\ 8 & 1 & 5 \end{bmatrix}$$

2 columns = 2 Rows
4 rows   3 Columns
Dimension of Product Matrix
4 × 3

This higher level of mathematics enabled statisticians to find complex mathematical solutions for statistical problems in a multivariate (or *p*-dimensional) space when a bivariate (or two-dimensional) system was insufficient.

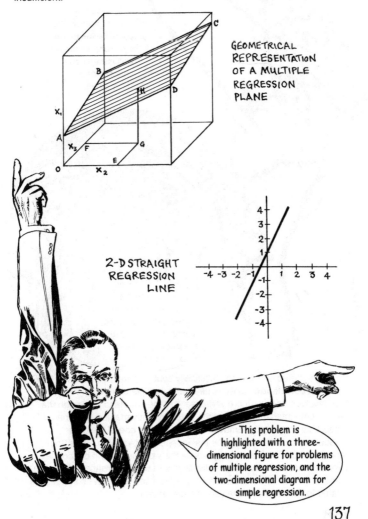

GEOMETRICAL REPRESENTATION OF A MULTIPLE REGRESSION PLANE

2-D STRAIGHT REGRESSION LINE

This problem is highlighted with a three-dimensional figure for problems of multiple regression, and the two-dimensional diagram for simple regression.

137

Statistical Control

Scientists can use two types of control when undertaking research: experimental and statistical control.

The former is produced by the action of the researcher through randomization and manipulation ...

... whereas statistical control, which involves mathematical manipulation, is often the first step towards experimental control.

Pearson offered one way to statistically control certain variables in 1895 with **part correlation**, which is used with multiple correlation only and thus involves three or more variables. It is the correlation between the dependent variable and one of the independent variables after the researcher statistically removes the influence of one of the other independent variables from the first independent variable. Thus, the researcher can mathematically isolate the variable when it cannot be experimentally isolated. The statistician is essentially treating the item as if one of the variables doesn't exist. (As we'll see later, part correlation is related to R.A. Fisher's analysis of covariance.)

For example, if dieticians wanted to determine which factors contributed to weight loss by assessing how important exercise, caloric consumption and fat intake were ...

... a multiple correlation might indicate that all three variables explained weight loss better than one variable alone.

But if the researchers wanted to look at the effect of reduction of calories alone, then they could use part correlation to remove the influence of fat intake and exercise from the combined set of independent variables. This would determine the role that caloric consumption had in weight loss alone.

George Udny Yule later introduced **partial correlation**, in which the statistician removes the effects of one or more of the independent variables from *both* the dependent and one of the other independent variables. Partial correlation helps to identify spurious correlations (see page 118).

Discrete 2 x 2 Relationships

Pearson introduced two new methods in 1900: the **tetrachoric** (i.e. "four-fold") correlation coefficient (r_t); and his **phi** coefficient (ϕ), known later as "Pearson's phi coefficient" for discrete variables. Both methods measure the association between two variables, designed for 2 x 2 (or four-fold) tables, which can be placed into two mutually exclusive categories, (called "dichotomous" variables).

	RECOVERED	DIED	
ESCAPED	a	b	
INCIDENCE			
INFECTED	c	d	

EXAMPLE OF A 4-FOLD TABLE USING AN EXAMPLE FROM PEARSON'S 1904 STUDY ON THE EFFICACY OF A VACCINE TO PROTECT AGAINST TYPHOID.

Pearson's phi coefficient was designed for two variables where a *true* dichotomy exists and thus the variables are not continuous. This technique is commonly used by psychometricians for test-construction in situations where a true dichotomy exists, such as "true" or "false" test items, and by epidemiologists who use it to assess a risk factor associated with the "presence" or "absence" of a disease against the incidence of mortality.

140

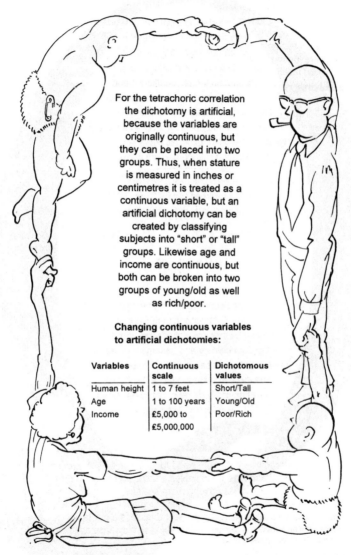

For the tetrachoric correlation the dichotomy is artificial, because the variables are originally continuous, but they can be placed into two groups. Thus, when stature is measured in inches or centimetres it is treated as a continuous variable, but an artificial dichotomy can be created by classifying subjects into "short" or "tall" groups. Likewise age and income are continuous, but both can be broken into two groups of young/old as well as rich/poor.

Changing continuous variables to artificial dichotomies:

Variables	Continuous scale	Dichotomous values
Human height	1 to 7 feet	Short/Tall
Age	1 to 100 years	Young/Old
Income	£5,000 to £5,000,000	Poor/Rich

Yule's Q Statistic

Yule proposed the Q statistic, which he named for Quetelet, in 1899 (one month after Pearson introduced the phi coefficient and tetrachoric correlation). Yule was also looking for a measure that didn't rely on continuous variables or depend on an underlying normal distribution, as was the case with the Pearson product-moment correlation.

I found that my Q (where the values ranged from –1.00 to +1.00) was always slightly higher than Pearson's tetrachoric correlation.

$$Q = \frac{ad - bc}{ad + bc}$$

Sociologists were among the first to use Yule's Q statistic in their work. It was adopted by medical statisticians at the end of the 20th century and became a measure of association for cases that could be obtained directly from cell counts in a 2 x 2 table, known as the odds ratio,* which is based on Yule's Q statistic.

George Udny Yule

*The odds ratio is a way of comparing whether the probability of a certain event is the same for two groups.

Biserial Correlations

Pearson devised the **biserial correlation** in 1909. This is related to the product-moment correlation (in which both variables are continuous), with one difference.

As we will see later, the biserial correlation is also similar to Student's t-test and Fisher's analysis of variance.

The **point-biserial correlation** is related to Pearson's biserial correlation, but one variable is continuous and the other is a "true dichotomy", such as male/female. This is an estimate of what the product-moment correlation would be if the dichotomous variable were replaced by a continuous variable instead.

These two methods are used quite extensively by psychometricians for test-item analysis in the construction of various intelligence and aptitude tests. The biserial is commonly used to determine the correlation between the score of a test item and the total scores on the test.

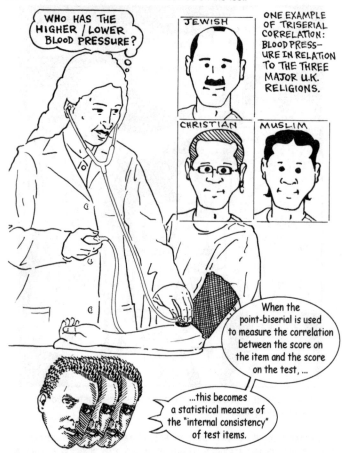

Pearson's triserial correlation is similar to the biserial, where one variable is continuous, but the second variable is a trichotomy (e.g., low, medium and high).

Egon Pearson and Polychoric Correlations

In 1922 Pearson and his son Egon devised the **polychoric** correlation. This is similar to the tetrachoric correlation, except that there are *three or more* possible values that a variable could have. While the tetrachoric is limited to a 2 x 2 contingency table with variables that are limited to binary values (0, 1) only, the polychoric correlation is used with an *n* x *n* table and the values of the variables are polyserial (0, 1, 2, 3, 4 …), and thus contain three or more categories.

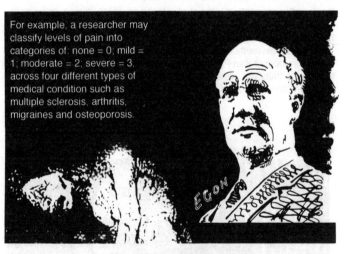

For example, a researcher may classify levels of pain into categories of: none = 0; mild = 1; moderate = 2; severe = 3, across four different types of medical condition such as multiple sclerosis, arthritis, migraines and osteoporosis.

EGON

LEVELS OF PAIN

Type of Disease	None = 0	Mild = 1	Moderate = 2	Severe= 3
Multiple Sclerosis				
Arthritis				
Migraines				
Osteoporosis				

Rank order correlation is the study of relationships between different rankings on the same set of items. It deals with measuring correspondence between two rankings, and assessing the statistical significance of this. Two of the main methods were devised by **Charles Spearman** (1863–1945, a student of Karl Pearson) and Maurice Kendall. Three other tests include the Wilcoxon signed-rank test, Mann-Whitney U test and the Kruskal-Wallis analysis of ranks.

I borrowed Galton's ideas of ranking values when I devised the Spearman rho (ρ) rank order correlation in 1906.

In principle, this method is simply a special case of the Pearson product-moment coefficient in which the data are converted to ranks, from the highest to the lowest, before calculating the coefficient.

Factor Analysis

Spearman was also influenced by Galton's ideas of measuring individual differences in human abilities and by his early ideas on intelligence testing. Using Pearson's product-moment correlation and the principal components method* that Pearson introduced in 1901, Spearman created a new statistical method, known as **factor analysis**, which reduces a set of complex data into a more manageable form that makes it possible to detect structures in the relationship between variables.

Spearman

I went on to create the first psychometric theory of intelligence with my two-factor theory, which measures general and specific abilities.

*A general statistical procedure for finding an efficient representation of a set of correlated data.

Maurice Kendall's Tau Coefficient

The English statistician **Maurice Kendall** (1907–83) created another ranking method of correlation in 1938, known as Kendall's tau. This method is a scheme based on the number of agreements or disagreements in ranked data.

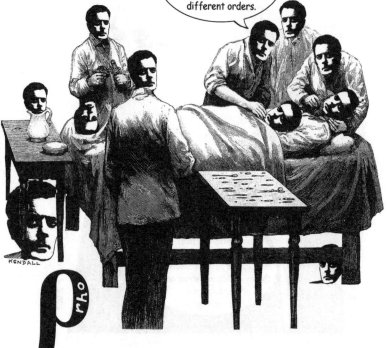

Kendall's tau is often used for samples larger than those used for the Spearman rho.

Correlation vs. Association

These terms are used to describe two different procedures to measure statistical relationships.

> I used the term <u>correlation</u> to describe the strength and direction of relationships between two or more continuous variables that assume a normal distribution.

> And in 1899 I introduced the term <u>association</u> for two or more discrete variables which do not assume a normal, continuous distribution.

Methods of **correlation** include:

Pearson's simple, multiple and part correlation,

biserial correlation,

triserial (and polyserial) correlation,

tetrachoric correlation

and **Yule's partial correlation**.

Measures of **association** where both variables are nominal:

the phi coefficient, **the chi-square statistic** (see pages 153-6), AND YULE'S Q.

Mixed measures when one variable is discrete and the second variable is continuous:

polychoric correlation, **Kendall's tau,** the **Spearman rho,**

the Wilcoxon signed-rank test, **the Mann-Whitney U test** and the **Kruskal-Wallis** analysis of ranks.

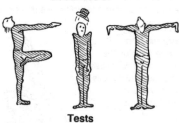

Goodness of

FIT

Tests

One of the ways in which the normal distribution is used to analyse or interpret data is by a method known as a *goodness of fit test*, which allows the statistician to see how well the data matches or corresponds to a normal distribution.

This means that the statistician can say whether the data is normally distributed, and can then make probabilistic statements about it.

Until 1900, this was the main way in which statisticians could say something about the probability of their results.

Our champion of the normal curve, Adolphe Quetelet, made one of the earliest attempts to fit a set of observation data to a normal curve around 1840, which Galton began to use in 1863. Quetelet's procedures were graphical, and he used a table based on the binomial distribution rather than taking an approximation of the normal curve. Much of Galton's work did not involve curve-fitting per se; instead, he compared his calculated values to a normal probability table.

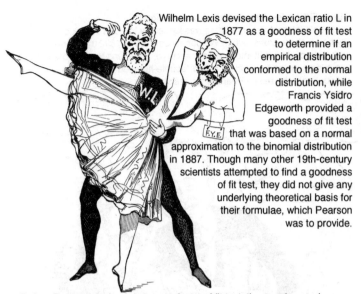

Wilhelm Lexis devised the Lexican ratio L in 1877 as a goodness of fit test to determine if an empirical distribution conformed to the normal distribution, while Francis Ysidro Edgeworth provided a goodness of fit test that was based on a normal approximation to the binomial distribution in 1887. Though many other 19th-century scientists attempted to find a goodness of fit test, they did not give any underlying theoretical basis for their formulae, which Pearson was to provide.

Before Pearson devised a new goodness of fit test, the usual procedure involved comparing errors of observation to a table of distributions based on the normal curve, or graphically by means of a plotted frequency diagram. Typically, as the evolutionary biologist **J.B.S. Haldane** (1892–1962) explained in 1936:

A researcher framed a scientific hypothesis and made an observation, and all that could be determined was that the two fitted very well or very badly ...

... but in intermediate cases there was no real test of the fit until Pearson devised his chi-square goodness of fit test.

151

Curve-Fitting for Asymmetrical Distributions

Pearson's interest in curve-fitting was fuelled by Weldon's work on Plymouth shore crabs. When Weldon discovered in 1892 that some of his crab data could not fit into one normal curve, and found that it produced two curves instead – what he called "double-humped" (or bimodal distribution) – he asked Pearson for assistance.

Pearson wanted to find another way to interpret the data without trying to normalize it as Quetelet and Galton had done. Pearson and Weldon thought it was important to make sense of the shape without distorting it, as it might have revealed something about the creation of new species.

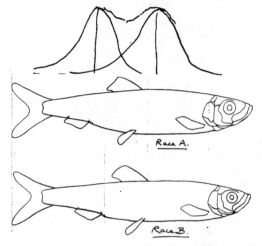

WELDON'S DRAWINGS OF DIFFERENCES IN HERRING WHEN HE & PEARSON WERE LOOKING FOR SIGNS OF SPECIATION.

Outlines traced from Heincke's figures of two typical Kiel Herring, one belonging to his Race A, the other to his Race B.

The Chi-square System

Pearson's ongoing work on curve-fitting throughout the 1890s signified that he needed a criterion to determine how good the fit was, which led him to devise different goodness of fit tests. By the end of 1896, he wanted to develop a goodness of fit test for asymmetrical distributions for biologists and economists, which culminated in his chi-square goodness of fit test in 1900.

There are three components to Pearson's chi-square (χ^2) system:

1. The chi-square probability distribution, published in 1900
2. The goodness of fit test of 1900
3. The chi-square test of association for contingency tables of 1904 (renamed the "chi-square statistic" in 1923 by R.A. Fisher)

But what was so important about the chi-square distribution and the chi-square goodness of fit test?

Their overriding significance was that statisticians could now use statistical methods that did not depend on the normal distribution to interpret their findings.

153

While the normal distribution is used for continuous data that conforms to the symmetric, bell-shaped curves, the chi-square distribution can be used for discrete data that takes on any-shaped distribution, such as asymmetrical, binomial or Poisson distributions.

Pearson's chi-square tests are built on two different hypotheses: the goodness of fit test determines how well an empirical distribution, set up by a researcher from data observed directly or from experimental data, could describe effectively the sample drawn from the given population (e.g. how well the experimental data fit the theoretical chi-square distribution).

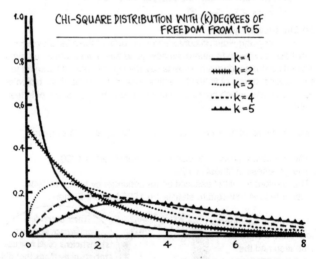

Conversely, the chi-square contingency coefficient, which is a measure of association, tests the difference between observed values and theoretically-expected values in a contingency table.

In the example below, a political analyst may want to determine if women or men are more inclined to vote Republican or Democrat in an American presidential election.

Voting preference in a 2 x 2 table:

Political Party	Gender		Total
	Female	Male	
Democrat	a	b	a+b
Republican	c	d	c+d
Total	**a+c**	**b+d**	**N**

The chi-square statistic for contingency tables can be best illustrated by a computational formula that Pearson devised in 1904 for 2 x 2 contingency tables where

$$\chi^2 = \Sigma \frac{n(ad - bc)^2}{(a + b)(c + d)(b + d)(a + c)}$$

The chi-square statistic might reveal that more women tend to vote Democrat while men tend to vote for Republican.

While the two chi-square tests perform different functions, they may be expressed mathematically (in contemporary terms) as:

$$\chi^2 = \Sigma \frac{(O - E)^2}{E}$$

or

chi-square =
$$\left(\frac{\text{the sum of all the values of (observed number – expected number)}^2}{\text{expected number}} \right)$$

The chi-square statistic is flexible and can handle more than one category, but here the more general formula would be used. Hence, in a British general election, where there are more than two parties, a political analyst may want to determine if women or men are more inclined to vote for the major political parties:

Observed values for voting preference in a 2 x 5 table:

Political Party	Gender	
	Female	Male
Labour		
Conservative		
Lib/Dems		
Green		
Nationalist Parties		

Interpreting Results with Degrees of Freedom
Unlike correlation, in which Pearson could look at the numbers (e.g. 0.90, 0.50 or 0.21) and know whether these values indicated high, moderate or low correlations, this is not possible with the chi-square statistic. One cannot look at the value derived from the formula and know what it means without additional assistance.

To interpret the computed chi-square values, Pearson devised what he called a "correction factor". R.A. Fisher formulated the "degrees of freedom" in 1922 to determine if the chi-square results were statistically significant or not. Degrees of freedom is based on the number of observations in the sample, and is universally used for most statistical methods.

Significance testing is an important concept because it allows the researcher to determine...

...whether the findings from a study are the result of a genuine difference or are due to chance.

THAT'S SURELY A DEGREE OF FREEDOM TOO FAR!

The Chi-Square Probability Table

Pearson and his student, Alice Lee (1858-1939), created a chi-square probability table in 1900. A year later, another student, William Palin Elderton (1877-1962), modified it. Having access to this probability table meant that the researcher could check the computed chi-square values and the requisite correction factor to determine if the results were statistically significant or not.

While Edgeworth discussed significance tests in 1885, Pearson's chi-square tests made it possible to determine the statistical significance of results on a larger scale than had been possible previously. Later generations of statisticians showed that there were other factors that could influence the correct degrees of freedom for the chi-square tests.

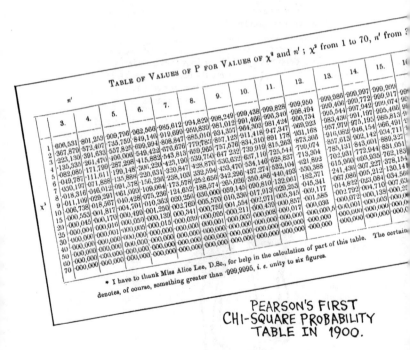

PEARSON'S FIRST
CHI-SQUARE PROBABILITY
TABLE IN 1900.

A Statistical Test for the Guinness Brewery

The first statistical quality control test for industry was devised by the statistician and chemist William Sealy Gosset, a master brewer at Guinness in the early years of the 20th century. Since Gosset was bound by his appointment not to publish under his own name (perhaps because Guinness didn't want rival breweries to know that they were training some of their scientific staff in statistical theory), he adopted the pseudonym "Student". This was standard practice at Guinness: Gosset's statistical assistant, Edward M. Somerfield, used "Alumnus" in his publications.

At the end of the 19th century Guinness was the largest brewery in the world, producing more than 1.5 million barrels a year.

To maintain this position, they began to appoint men who had first-class science degrees from Oxford and Cambridge, and they also adopted a policy of sending staff away for specialized study.

Quantifying Brewery Material

Guinness also had large farming interests, especially in growing barley for beer, which led to Gosset's involvement with agricultural experiments and laboratory tests.

Some of the brewers undertook chemical analyses to try to identify and quantify the qualities that made hops and barley good brewing material ...

... such as the "rub" of the hops or the "texture" of barley that might be "milky" or "steely".

But these qualitative criteria were difficult to measure, so Guinness didn't know precisely what it was that made their stout so popular, or how to improve or maintain the quality. They wanted to know what conditions were needed for producing those varieties of barley and hops that gave the best malting and brewing value respectively.

160

Agricultural Variation

When Gosset arrived at Guinness and saw the immense amount of chemical data at the brewery, he wondered if this information could show the relationship between the quality of raw material, such as the barley and hops, and the quality of the finished product. Two difficulties Gosset encountered, when he began to plan statistical analyses, were that the variation was high and the statistical observations were few.

Variation in rainfall, bird damage, soil chemistry and temperature were critical in the agricultural production of these cereals, but the brewers didn't know how to take account of this variation when interpreting the data.

Guinness thus needed to find a way to decide which differences to ignore and which to take seriously. One way to analyse the variation was by using Pearson's statistical methods. Gosset arranged to meet Pearson on 12 July 1905 at East Ilsley in Berkshire, where Pearson spent his summer vacations so he could be within cycling distance of Weldon in Oxford.

Small Samples vs. Large Samples

Gosset told Pearson that one of his biggest problems was the small sample sizes – he had a sample size of ten for each variety of barley. This was, indeed, very small compared to the large samples that Pearson used elsewhere: it was this problem that led Gosset to devise the first statistical quality control test.

Gosset adapted Pearson's methods for smaller sample sizes and also borrowed some of the statistical methods used by astronomers. These combined linear equations of observations were, however, of rather limited use because they were concerned with observations made under *stable* conditions, whereas the agricultural conditions of the brewing data were *unstable* – they were highly variable and were also affected by changes made in laboratory experiments.

Testing Statistical Differences Between Two Means

Through combining the astronomers' methods with Pearson's statistical methods, Gosset created the statistical tools he needed for his experimental data. He wanted to find out if there was a significant difference between two fertilizing treatments used for two varieties of barley on neighbouring plots, which were subject to different types of soil, manure and weather conditions.

Which variety would produce the better quality in terms of value per acre?

The idea of calculating differences between two group means, and trying to find a meaningful way to interpret the data for small samples, had already been considered by French physician Pierre Louis and German physicist Gustav Radicke in the 1850s, though they had not been successful. Gosset introduced his **z-ratio** (or z-test) to determine if there was a significant difference between the sample mean and the population mean.

163

Statistical Results for Guinness

When Gosset analysed the barley plots using his newly devised z-ratio, he found that the best barley for Guinness was the Archer variety. Once Guinness knew which barley to buy, they wanted to grow it all over Ireland.

Student's z-ratio became the first statistical test for quality control in industry. Gosset's ideas, which demonstrated that it was important to determine the quality control of a product, influenced a new generation of statisticians, including R.A. Fisher, Walter Shewart (1891-1967) and W. Edwards Deming (1900-93).

Student's t-test

Fisher was so inspired by Gosset's statistical test that in 1924 he transformed Gosset's z-ratio and re-introduced it as "Student's t-test". Fisher then recalculated Gosset's values from the z-table and replaced it with a t-table, which Fisher termed "Student's t distribution". Student's t-test was re-expressed as:

$$t = \frac{\text{sample mean from Group 1 − sample mean from Group 2}}{\text{Standard error of the differences}} \quad \left\{ \frac{\bar{x}_1 - \bar{x}_2}{se} \right\}_{\substack{\text{standard} \\ \text{error}}}$$

There are three different ways in which the t-test is used:

It tests the mean differences between two independent samples.

It tests the mean differences between two matched samples.

It is a test for regression coefficients.

Fisher further developed Gosset's work when he devised his "analysis of variance" for his classic design of experiments on the Broadbalk wheat data at Rothamsted Experimental Station, Harpenden, Hertfordshire (north of London).

A New Statistical Era: Rothamsted's Broadbalk Agricultural Data

Though Pearson offered Fisher a post at University College London in 1919, Fisher decided instead to take up Sir John Russell's offer to work at the Rothamsted Experimental Station to analyse the Broadbalk agricultural data, where his statistical innovations came to fruition.

Rothamsted is one of the oldest agricultural centres in the world, established in 1834 by **John Bennet Lawes** (1814–1902) whose ancestors had owned the land since 1623.

After taking his degree at Oxford, Lawes returned to Rothamsted Manor and turned his barn into a chemical laboratory where he undertook experiments on mineral phosphates with different quantities of sulphuric and other acids.

This formed the beginnings of the artificial fertilizer industry and revolutionized British agriculture.

In 1834 the chemist **Joseph Henry Gilbert** (1817–1901) joined Lawes in his work on the experimental cultivation in the Broadbalk fields. From their work, they published all the statistical details of their observations and experiments, and found that the continuously fertilized plots produced 12 to 13 bushels a year while the well-manured plots yielded 30 to 40 bushels a year.

The end of the First World War in 1918 brought a period of expansion and reconstruction at Rothamsted. In the following year the agricultural chemist **Edward John Russell** (1872–1965) hired the Cambridge-trained mathematician Fisher.

HARVESTING BROADBALK

Russell

I was asked to stay as long as was necessary to determine if their records were suitable for a proper statistical analysis.

Fisher's Statistical Analysis of Variance

From 1919 to 1926, Fisher pioneered the principles of the design of experiments and further developed his statistical methodology of the **analysis of variance** (ANOVA) that he had begun in 1916. While all experiments are concerned with the relations between variables, there had been no systemic way of assessing these relationships until Fisher provided a novel and highly innovative methodology in his influential book, *Statistical Methods for Research Workers* (1925).

At Rothamsted, Fisher's task was to analyse statistically the data of weather, crop yields and fertilizers that had accumulated over a period of 66 years.

I decided to look at the amount of variation in the data to determine which factors were influencing the quality of wheat.

The Analysis of Agricultural Variation

Fisher realized that it was essential to distinguish between three types of variation in the wheat yield: *annual variation*, which was due to the direct effect of the meteorological conditions stimulating plant growth, as well as the physical effect of the soil; *steady variation* that was attributed to the deterioration of the nutrients held in the soil; and *slow changes* in variation, which were small unexpected changes.

By analysing the different wheat plots, I discerned from the overall picture (i.e. the main effect) that the influence of excessive rain on the crops was removing soluble nitrates.

His findings suggested that nitrogen-rich fertilizer should be used in the spring rather than the autumn to improve the quality of wheat.

The Analysis of Variance and Small Samples

The analysis of variance is a methodology consisting of a collection of statistical models for experimental data that divides the observed variation into different parts; this partitioning of variance is integral to Fisher's statistical methodology.

While Student's t-test is used to determine if there is a statistically significant difference between two group means, ...

...Fisher's analysis of variance uses an F-test followed by the F-table to determine if there is a significant difference in the group mean.

If the value is significant, t-tests are used to test differences between any two means to locate where the differences lie.

Fisher introduced the **analysis of covariance** (ANCOVA) in 1932 to statistically control a variable. That is, one "covaries" out the influence of one variable from the other variables, which may increase the precision of the experiment by reducing the error variance. Pearson provided a similar analysis with part correlation in 1896.

Inferential Statistics

Building on Pearson's methods, Fisher not only expanded the existing terminology, but his statistical innovations inaugurated the second phase of modern mathematical statistics through his development of **inferential statistics**. While random variation is the basis for inferential statistics, the distinctive feature of this newer form of statistics involves the formal testing of hypotheses and estimation theory.

Hypothesis testing is a scientific procedure for making rational decisions about two different claims. *Estimation theory* is a branch of statistics that deals with estimating the values of parameters (see next page), based on data collected by the scientist. For example, if a political analyst wants to estimate the proportion of the population of voters in the UK, the proportion is the unknown parameter and the estimate is based on a small random sample of voters.

Statistics, which use Roman letters such as $\bar{X}$, s, and r (for the mean, standard deviation and correlation, respectively), were developed primarily by Pearson.

Parameters, denoted instead by Greek letters such as μ (mu), σ ("little sigma") and ρ (rho), were introduced by Fisher in 1922 to estimate the mean, standard deviation and correlation, respectively, in populations.

Hence, statistics are to samples what parameters are to populations.

The Sampling Distribution

To make generalizations about a population, statistical information is taken from a representative sample.

Every sample drawn from the population has its own statistic ($\bar{x}$, s, or r), which is used to estimate the parameter (μ, σ, or ρ) of its population. According to Fisher, a sample statistic should be an unbiased estimator of the corresponding population parameter. (Fisher created three other estimators for parameters, which had to be statistically consistent, efficient and sufficient.)

To get from a sample statistic to an estimate of the population parameter, a statistician uses a "sampling distribution". Instead of using one sample, several samples (or an infinite number of samples) are drawn from the population; each sample would produce a different mean, standard deviation and correlation. The average of these statistics from all the samples should be close to the population average.

Hence, a population parameter is a way of summarizing a probability distribution, while a sample statistic is a way of summarizing a sample of observations.

The foundations of Fisher's method were not only built upon Pearson's statistical work, but represented a translation of Pearson's statistical language. They became the vernacular of contemporary mathematical-statistical theory, even though many of Pearson's statistical methods and his language remain a part of statistical theory.

Conclusion

The bureaucratic compilation of the vast amount of vital statistical data by mid-Victorian statisticians enabled them to deploy a statistical system to measure the health of the nation, which led to political reform and the creation of Public Health Acts in Britain. The vital statisticians' idea that statistical variation was flawed and a source of error to be eradicated was challenged by Charles Darwin's ideas of biological variation and of statistical populations of species. This Darwinian framework prompted a reconceptualization of a new statistical methodology beginning with Francis Galton, whose interest in measuring individual differences brought variation into the forefront of statistics; his work captivated the interest of W.F.R. Weldon, whose ideas, encouragement and support provided the impetus for Karl Pearson and his colleagues to establish the foundations of modern mathematical statistics.

The first statistical quality control test for industry was devised by Pearson's student William Sealy Gosset, whose work inspired Ronald Fisher to create a statistical system for the analysis of small samples, thereby introducing experimental design and randomization into statistical theory. Fisher's development of inferential statistics inaugurated the second phase in the development of modern mathematical statistics.

Ever since the 20th century, statistics has become the language for medical, economic and political dialogues. Consequently, it has infiltrated the vernacular of daily conversation. Statistical information can wield powerful influence on people's lives, affecting decisions about medical treatment, which car or house to buy, which clothes to buy and which political party to support in an election. In the technologically-driven Information Age of the 21st century, an understanding of statistics remains paramount to our lives.

Bibliography

Bowler, Peter (2003), *Evolution: The History of an Idea*, 3rd edition, Berkeley & Los Angeles, CA: University of California Press

Gigerenzer, Gerd (2002), *Reckoning with Risk: Learning to Live with Uncertainty*, London: Penguin; US edition: *Calculated Risks: How to Know When Numbers Deceive You*, New York: Simon & Schuster

Goldacre, Ben (2008), *Bad Science*, London: 4th Estate

Gould, Stephen Jay (1996, 1981), *The Mismeasure of Man*, New York: WW. Norton

Gould, Stephen Jay (1996), *Full House: The Spread of Excellence from Plato to Darwin*, London & New York: Three Rivers Press

Hacking, Ian (1990), *The Taming of Chance*, Cambridge: Cambridge University Press

Hacking, Ian (2006), *The Emergence of Probability*, Cambridge: Cambridge University Press

Huff, Darrell (1996, 1954), *How to Lie with Statistics*, New York: Norton & Company

Porter, Theodore M. (1986), *The Rise of Statistical Thinking, 1820–1900*, Princeton, NJ: Princeton University Press

Sardar, Ziauddin, Jerry Ravetz and Borin Van Loon (2006), *Introducing Mathematics*, Cambridge: Icon Books

Stigler, Steven M. (1986), *The History of Statistics: The Measurement of Uncertainty Before 1900*, Cambridge, MA: Belknap Press of Harvard University Press

Stigler, Stephen M. (1999), *Statistics on the Table*, Cambridge, MA: Harvard University Press

The author
Eileen Magnello trained as a statistician before doing her doctorate in the history of science at St Antony's College, Oxford University. She has published extensively on Karl Pearson and co-edited *The Road to Medical Statistics*.

The illustrator
Borin Van Loon is a surrealist painter and illustrator. His collage comic strips are compiled in *The Bart Dickon Omnibus*, a graphic novella. Having worked on *Darwin*, *Buddha* and *Cultural Studies*, this is the fifteenth documentary comic book he has illustrated and designed.

Index